The First Arkansas Union Cavalry 1862-1865

by
Russell Mahan

2nd Edition 2019
Published by Historical Enterprises
Santa Clara, Utah
HistoricalEnterprises@Outlook.com

Published by Historical Enterprises
Santa Clara, Utah
Copyright © 1996, 2019 by Russell Mahan

No part of this book may be reproduced, scanned, or distributed in any printed or electronic form without permission.

First Edition 1996
Second Edition January 2019
Printed in the United States of America
ISBN 9780999396254

On the front cover of this book is a photograph of Lieutenant Thomas Wilhite, courtesy of the Prairie Grove Battlefield State Park.

The First Arkansas Union Cavalry 1862-1865

Table of Contents

The Persistence of Arkansas Unionism. 1

Humiliation at the Battle of Prairie Grove.. 9

A Cruel Civil War Among Neighbors. 19

Vindication at the Battle of Fayetteville. 29

Life in the First Arkansas Union Cavalry. 53

Hard Duty in the Saddle. 71

The Post Colony System. 89

Peace from the East. 95

Endnotes.. 99

Selected Bibliography. 103

Index of Members of the 1st Arkansas Cavalry. . . 105

Chapter 1
The Persistence of Arkansas Unionism

On the afternoon of Monday, May 6, 1861, an intense political excitement was sweeping Arkansas. The moderation which had prevailed in the councils of government and in the hearts of most of the people was blown away in the cannon fire at far away Fort Sumter, and replaced by the virtual unanimity which only war can produce. The very men who had voted to remain in the Union just a few weeks before at the first session of the Arkansas Secession Convention were now reconsidering.

Those same convention delegates were gathered for a second session at the chambers of the State House of Representatives in Little Rock. In its original voting in early March, the convention had voted down three secession resolutions by votes of approximately 40 to 35. Now, recalled to the task in the wake of war in South Carolina, disunion was resisted by only five delegates. David Walker of Fayetteville, president of the convention and previously a Unionist himself, called for a unanimous vote for secession.

The political pressure on the five remaining Union delegates was unbearable. Standing in the path of a tidal wave of enthusiasm was more than most men could do, especially when to do so called forth the bitterest denunciations of treason from erstwhile friends and associates. It was no wonder that four of the five opponents changed their votes to favor the secession they could not stop.

Then all eyes turned upon the last dissenter, a delegate from northwest Arkansas, Huntsville educator Isaac Murphy. He rose to his feet to speak. "I have cast my vote after mature reflection, and have duly considered the consequences," he boldly explained, "and I cannot conscientiously change it. I therefore vote 'No.'" Amid a storm of verbal abuse and scorn which instantly burst from the other representatives and the public in attendance, a woman in the gallery threw a bouquet of flowers to Arkansas' last Unionist delegate. It was among such people as these - a man with his vote and a woman with her flowers - that the First Arkansas Union Cavalry had its beginning.[1]

There were in fact many Union people in Arkansas, particularly in the northwest part of the state. They were a minority to be sure, but nonetheless a significant one. In Washington County it is estimated that in the course of the war 2,000 men served in Confederate army, and 500 to 800 in the Union.[2]

In the months following secession many Union men had experiences similar to a young man by the name of Thomas Wilhite. In June of 1861, about a month after the secession of Arkansas, six men rode up to the Wilhite farmhouse at Fall Creek, Arkansas, about 21 miles south of Fayetteville. Wilhite, just 24 years old, saw them coming and armed himself with a rifle and revolvers.

For some time Wilhite had by necessity been extremely cautious. As an unabashed Union man in a Confederate state, he had learned to keep a careful watch to his rear. Working in the fields, he kept a musket slung over his shoulder. Sleeping at night, he not only kept a rifle at the head of his bed but sometimes wore pistols around his waist as well. Men had occasionally come to arrest him at his own farm, but finding the plow without its owner, they would leave the equipment unmolested in the eery belief that they were at that moment within the deadly aim of the unseen Wilhite.

Upon the arrival of his six visitors, Wilhite came around the corner of the house, taking them by surprise. They informed him that they had come to arrest him for being a Union man. Holding his musket on them, Wilhite replied that they had better get reinforcements for the job as six men was not enough. He cautioned them against going for their guns, as he would take at least one man with him for sure, and there was no telling just which man that would be. The visitors reconsidered the situation, then asked that they be permitted to leave unharmed. Wilhite generously granted their request.

On a different occasion, another group of men came to the farm. Finding their intended victim away from home, they consoled themselves by showing their hanging rope to Wilhite's hapless mother. One day Thomas Wilhite went over to a mill owned by T.K. Kidd near Cane Hill on a business errand. He found himself surrounded by several men who had the usual objective of arresting him. Armed, as always, Wilhite told them he had 13 shots, and that he would not be taken alive. The crowd mellowed at this point, and disbanded when Mr. Kidd

Thomas Wilhite as a Lieutenant in the First Arkansas Union Cavalry. (Prairie Grove Battlefield State Park).

suggested that Wilhite might yet make a good Southern soldier. This would turn out to be true, but not in the way Mr. Kidd intended.

Attending a Baptist gathering one Sunday, Pastor Thomas Dodson launched into a torrent of abuse against Union sympathizers. "If there is a Union man within the sound of my voice," he declared, "I want him to leave the house, and leave it now-a." Wilhite, in his unrestrained and unintimidated way, headed for the door. "Then go-a," the minister intoned, "and darken not again the house of God."

By November of 1861 Wilhite had all that he could take. He felt compelled to go into hiding, taking refuge in a cave in the Boston Mountains south of Fayetteville which he had previously supplied with corn and forage for just such an eventuality. There he stayed with several horses, only occasionally returning home, and then with the greatest of caution.[3]

The Union victory at the battle of Pea Ridge, Arkansas, in March of 1862 considerably dampened Confederate aspirations in Missouri and encouraged the many Unionists in northwest Arkansas. Such men began to work their way northward to Federal Army lines. By May Thomas Wilhite was himself ready to take up arms in defense of his nation and against the majority of his home state, and he set out for the Union lines in southwestern Missouri. By the time he arrived, he had thirty men with him. They would be back - armed, mounted and wearing blue.

Lieutenant Colonel Albert W. Bishop would come to serve with these Union fugitives such as Thomas Wilhite and to know their personal stories well. In his partisan way, he wrote in late 1862:

> The Government knows but little of the sufferings of the loyal men of the Border. It is no easy thing to adhere to the Union in a seceded State, and when insult, outrage and beggary are the consequences..... In no section of the country has the Great Rebellion created such intense personal hate, or separated more widely friends and relations, than in the South-West....
> The poison [of disunionism] spread, and soon infused itself into the minds of hundreds of peaceable citizens, transforming them into bands of armed and head-strong men, ready at a moment's notice to fire the house, plunder the property, and take the life of an inoffending neighbor, if suspected, even, of sympathy with the "Lincoln Government." Nobody, in fact, could be so bad as a "Fed".... Personal abuse was followed up by the shotgun....
> To remain longer at home was worse than to leave wives and children (temporarily, as they thought) and thus began the hejira of the Southwest. About this time Federal forces were again accumulating at Springfield, and thither hunted, but not disheartened, Unionists of Arkansas bent their steps.... No obstacles daunted, no dangers appalled them. Lying in the woods by day, at early nightfall they resumed their toilsome journey, carefully shunning highways and trusting to the instinct of self-preservation...for ultimate safety....[4]

In the spring of 1862, the refugees began to appear at the Federal military post in the southern Missouri town of Cassville. Captain Marcus LaRue Harrison of the 36th Illinois Infantry was the quartermaster of the Federal garrison at Cassville. He liked these Arkansas men who had given up their homes for the Union. When he was authorized to raise a company for the 6th Missouri Cavalry, he decided to fill the enlisted ranks with them. When the quota was quickly over subscribed, the idea emerged to form an entire regiment from this pool of manpower.

Many Yankee officers did not agree with Harrison's enthusiasm for these men. A regiment of Union men from seceded Arkansas was, many Northerners thought, just a plain bad idea. The belief was commonly held that such men necessarily had divided loyalties and in the final analysis would not fight as soldiers must. "There are officers in the army," Bishop wrote later, "who knowingly shook their heads at the project, and prophesied nothing but failure."[5]

Colonel Marcus LaRue Harrison was 32 years old when he organized the First Arkansas Union Cavalry. He was the son of a New York Presbyterian minister, and had attended Yale to study theology to be a minister himself. He instead became an engineer and worked in the rail-road business in Illinois and Iowa. When the Civil War began he joined an Illinois regiment and was sent to Missouri. (Shiloh Museum.)

Yet others felt quite differently. Brigadier General Egbert B. Brown, the commander of the military District of Southwest Missouri, and Missouri Military Governor John S. Phelps were instrumental in getting the regiment approved in Washington, D.C. Finally, on June 16, 1862, the War Department issued a special order to Captain Harrison, stating that "the Secretary of War hereby authorizes you to raise a regiment of cavalry from the loyal men of Arkansas, to be completed by the 20th of July, and to be mustered into service, clothed, mounted, and armed at Springfield, Missouri, by the United States government. The regiment will be mustered into service for three years or the war...."[6]

Meanwhile, Arkansas men kept coming into Union lines. On May 10th, eleven men led by Thomas J. Gilstrap came to the Federal picket line. Four days later was the arrival of Thomas Wilhite and his thirty men. Then on June 20th, a hundred and fifty men from Washington County rode in under Thomas J. Hunt, a young teacher from the Fayetteville area. There was no doubting that Arkansas Unionism was on the rise.

Staffing the new regiment was a big project. Officers and men alike came from many sources and backgrounds. Colonel Harrison did not go into the regiment alone, but took with him his family and friends. His younger brother Elizur, age 22, entered as a private but soon found himself a Lieutenant. Also, the Colonel's son, Edward M. Harrison, enlisted. Although only 11 or 12 years old, he was registered as 18 and designated a bugler. Two of Harrison's compatriots from the 36th Illinois Infantry, Sergeant James J. Johnson and Quartermaster Sergeant James Roseman, were brought on board as First Lieutenants of Companies A and M, respectively.

The position of second in command of the regiment fell to Captain Albert W. Bishop of the 2nd Wisconsin Cavalry, who was promoted to Lieutenant Colonel. He brought with him that regiment's chief trumpeter, Albert Pearson, who was made Second Lieutenant of Company K.

Lieutenant Colonel Albert W. Bishop was 30 years old when he joined the First Arkansas Union Cavalry. He had been educated at Yale and started a family and a law practice in his native New York. In 1860 his wife died and he moved to Wisconsin. When the war began he became a lieutenant in the artillery and later a captain of cavalry. During the war he wrote the book Loyalty on the Frontier *about the members and experiences of the First Arkansas Cavalry. (Washington County Historical Society.)*

Other commissioned officers for the new regiment came from two sources. First, authorizations to raise troops were given to local men from southwest Missouri and northwest Arkansas who were thought sufficiently eminent in their home districts to recruit a number of men. Sergeant John I. Worthington of the 6th Kansas Cavalry, formerly of Carroll County, Arkansas and presently of Granby, Missouri, was authorized to raise a company. Upon doing so, he became Captain of Company H. Civilian Charles Galloway of Barry County, Missouri, brought in many recruits and was appointed Captain of Company E.

Thomas Wilhite of Washington County, Arkansas, was rewarded with the rank of Lieutenant for his recruitment success.

A second source of commissioned officers was the lower ranks of existing Federal units. The formation of the First Arkansas Union Cavalry created a great opportunity for promotion for many enlisted men of other regiments. Transferees included Private James H. Wilson of the 37th Illinois Infantry (promoted to First Lieutenant of Company D), Corporal Jacob J. Reel of the 4th Iowa Infantry (First Lieutenant of Company I), Corporal Henry W. Gildemeister of the 1st Missouri Cavalry (Second Lieutenant of Company I), Private John Bonine of the 4th Iowa Infantry (Captain of Company L), and Private Joseph S. Robb of the 4th Iowa Infantry (First Lieutenant of Company L). There were others as well.

The recruiting of enlisted men also went on in two areas - in Missouri from Missourians and the walk-in refugees fleeing from Arkansas, and in Arkansas among Union men still at home. W.C. Peerson of Company B illustrated the former situation. He wrote near the end of the war of his motives for enlisting:

> *I was borned in Washington County on the 23rd of March 1844. I lived with my Parance until about the age of 18. In the year 1862 (the most of that time we lived in the State of Arkansas) I was forced by the Rebels to leave my home and take sides with them or go North to the Federal Army and I took it my choice to go North - myself, my Father, other relations and acquaintances 8 in number. Started on the morning of the 1st of December 1862 for the Federal Army, of which the nearest point was Elk Horn in Benton County, a distance of 35 miles. There was some Rebels between where we started from and the Federal Army although we had no trouble on the route. We reached the Army on the night of the same day, where we remained until the 3rd of December when myself and friend Campbell joined Co. B, 1st Arkansas Cavalry Volunteers on the morning of the 4th.*

Recruiting efforts went on secretly in northwest Arkansas. On June 27, 1862, General Brown reported "Carroll, Madison, Benton, and Washington Counties have been thoroughly scouted..., the expedition bringing in about 100 recruits for the First Arkansas Regiment.

Northwest Arkansas is reported loyal, and its permanent occupation would demonstrate it."

The primary means of recruitment was for word of a clandestine meeting in a secluded location to be whispered among known Unionist sympathizers. There was danger both in the telling of the meeting and in the attending of it. "Recruiting in Arkansas for the Union Army was at that time a perilous undertaking," wrote Lt. Col. Bishop. "Loyal men avowed their principles at the hazard of life, and the greatest difficulty was in getting recruits to the rendezvous of the regiment for which enlistments were being made."

"From the enlistment of its first man," Bishop continued, "to the mustering of the twelfth company, the camp of the regiment was a continuous story of wrongs and outrages, and old men and boys, women and children, were subsisted by the Government, whilst husbands and brothers were preparing for the avenging strife. Singly and in groups they came to Springfield. Weary and sore, they stood up to be 'sworn in,' many infirm of limb, but firm of purpose, and thus arose the regiment."

Chapter 2
Humiliation at the Battle of Prairie Grove

The regiment was scarcely organized when the seriousness of its purpose was brought home. James Daniels of Benton County, Arkansas, enlisted mustered into the service on July 3rd, and his military career lasted exactly one week. On July 10th he was back in his home county when Rebel guerrillas found and killed him. Private Daniels was the first man in the First Arkansas Union Cavalry to be killed by the enemy.

In late August of 1862 a portion of the regiment was assigned to strike into northwest Arkansas. This was surely a thrilling moment for the members of the regiment who participated. They were going home, not on a recruiting expedition, but as full-fledged Union cavalry on scout. Captain Charles Galloway was in command, with Captain John I. Worthington second in command. Riding first to Berryville and then on to Carrollton, they encountered a sizeable "guerrilla band strongly posted on a bluff, a short distance east of town, and partially concealed by a rude breastwork of logs. They evidently intended to 'stand,' and seemed to think themselves secure against cavalry." However, when the Federal cavalry opened fire with their Whitney rifled muskets, the Confederates discovered that their inferior weapons were no match in range. Turning to a quick retreat, all but two of the Rebels escaped from the charging cavalry. There would be many such scouts.

The random infliction of death on members of the regiment began in earnest and would continue for nearly three years. Private Ahaz Dunfield of Company H, a 26-year-old man who had enlisted 32 days before, was "killed by the enemy carrying the mail, Sept 17 62 between Elkhorn Ark & Cassville, Mo."

It was undoubtedly the hope of all members of the new regiment that they advance into Arkansas to reclaim their state and their homes for the Union. Instead, a discouraging retreat occurred. Resurgent Confederates gained control of Cassville and southwestern Missouri and Federals withdrew northward toward the interior of the state. "The abandonment of Cassville," Lt. Col. Bishop wrote, "was a serious blow to the buoyancy of the Arkansas men, nearly a thousand of whom were now at Springfield. Cassville was fifty miles nearer their homes, and they regarded that post as an indispensable link in the chain of communication that would ultimately re-unite their native hills and valleys to fatherland...."

On September 21, 1862, about 175 men of the First Arkansas Union Cavalry made a raid on the Confederates at Cassville catching them completely by surprise. In a two-pronged attack, Captain Worthington struck from the south and Captain Jesse Gilstrap from the north. The Rebels engaged in some "excellent skedaddling," but before Rebel reinforcements could arrive, the Federals had themselves vanished from town. The attack was an adventure for everyone in the regiment who participated, except 18-year-old Private Hiram Ross of Company E, who lost his life in the skirmish.

In the give-and-take nature of the Civil War Cassville was very soon back in Federal hands as Confederates fell back southward. Better yet, on October 18, 1862, the First Arkansas Union Cavalry was ordered to set up a post in Arkansas itself, at Elkhorn Tavern. That was the very site of the Federal victory at the battle of Pea Ridge, which had taken place the previous March. Undoubtedly a wave of excitement swept through the men upon this announcement. Going back to re-take Arkansas was exactly the purpose for which they had enlisted.

The assigned duty of the regiment at the advance Union outpost of Elkhorn Tavern was a difficult one. A letter from General James M. Schofield was received which set forth this virtually impossible order:

> *That no misapprehension may exist, this is to inform you that your forces are expected to continually scout and scour all the country within your reach. One-half of the command may be on distant scouts all the time; the other portion should be constantly employed in your immediate neighborhood. No part of your forces should be idle at any time. You are expected to rid all the country within your reach of all small bands, guerrillas, provost guards, etc., etc. Your forces should continually harass the enemy by driving in pickets and skirmishing with advanced guards and detached parties, capturing forage trains and commissary wagons.*
> *No limit is placed upon the country through which you may act, but you are expected to go wherever*

you can, without necessarily jeopardizing your command. You are to relieve the Union people and punish the treasonable. Unfailing activity and the utmost vigilance are demanded at your hands. One large party, consisting of about one-half your men, should be pushed near the enemy's lines and kept out all the time, capturing pickets, etc., and you may even go in the rear of the enemy's forces, and do them all the damage you possibly can. Feel the enemy often, and communicate all information you may obtain. This force should be relieved by the other half after a scout of five or six days.[7]

Lt. Col. Bishop, who was in command at Elkhorn while Colonel Harrison was on detached duty in Missouri, was incredulous. "All this," he marveled, "was expected from two battalions of cavalry [the third battalion was still organizing], who had never been one hour in a camp of instruction; and though now in the service from eight to nine months...had...been only partially clothed - there was not an overcoat in the line - and has never been paid.... But the men knew the country where they were operating. They were in their native hills again, and were active and zealous in the efforts to support that Government, loyalty to which had caused them so much suffering."

In mid-November of 1862 Companies G and K under Captains Rowen Mack and Theodore Youngblood were sent into Carroll County, Arkansas, to escort a number of fleeing Union families from there to Missouri for protection against retaliation. They ran into what was estimated to be several hundred Confederates and a brief but sharp fire fight at Yokum Creek on November 15th. Private Samuel Littrell was shot through the left arm, breaking it. Private James Standlee received a flesh wound in the left hip, and Private Luther Phillips of Carroll County was killed. Many other scouts were also sent out.

Meanwhile, larger events were occurring. Union General James G. Blunt and Confederate General Thomas C. Hindman were both spoiling for a confrontation, and as December came on, it was clear that the moment of decision was close at hand. The First Arkansas Union Cavalry was about to have its first real battle.

Colonel Harrison and the regiment left Elkhorn Tavern on December 5, 1862, as part of the southward movement of General Francis J. Herron's Union army. Herron was hurriedly moving to join with the further advanced General Blunt before Hindman's Confederate forces could strike. On the evening of the next day the First Arkansas arrived at the Illinois River southwest of Fayetteville and encamped. Colonel Harrison sent a message to General Blunt, who was at Cane Hill, just a few miles away, stating that his men and horses were too tired to proceed further, and that he did not think he could move them before Monday morning.

Blunt was furious with this message. It was Saturday night, Confederates were about to attack him, and when within only a few miles of helping them, Harrison sent a note like this. Harrison's men had come a shorter distance on horses than General Herron's men had come on foot, and the infantry was still coming. In his report on the Battle of Prairie Grove, Blunt justifiably complained:

> About 9 p.m. of the 6th, I received a note from Colonel Harrison, of the First Arkansas Union Cavalry, who had been ordered down from Elkhorn at the same time that General Herron started from Wilson's Creek, informing me that he had arrived at Illinois Creek, 8 miles north of Cane Hill, with five hundred men, and that his horses and men were so tired that he did not think he could move farther until Monday, the 8th.
> Whether his regard for the Sabbath or the fear of getting into a fight prompted him to make such a report to me, I am unable to say; but, judging from his movements that he was not a man upon whom to place much reliance on the battle-field, I ordered him to proceed by daybreak to Rhea's Mills, to guard the transportation and supply trains....[8]

Unfortunately, this note by Harrison remains a mystery. In the regimental records available today, no comment is made by him about it. Bishop in his writings did not comment either, which may have been because he was not present (he remained in command at Elkhorn Tavern), or perhaps because he did not wish to criticize the commander under whom he served at the time of writing and publishing. Clearly, however, it was a low point for M. La Rue Harrison personally. Unfortunately, the regiment as a whole was going to hit a low the following day.

On the early morning of December 7, 1862, Confederate General Hindman put in motion one of those command decisions which, if it works, is brilliant, and if it doesn't, is foolish. He decided to march quickly between the converging armies of Union Generals Blunt and Herron, strike and defeat Herron before they could unite, then turn and defeat Blunt. That morning, between Herron and Blunt were a few regiments of Union cavalry, which were the vanguard of Herron's army.

The lead of these regiments was the 7th Missouri Union Cavalry, behind which were the 8th Missouri and then the 1st Arkansas Union Cavalry. The unsuspecting 7th Missouri was taking a rest from their fatiguing relief march when disaster struck. The last word from Blunt had been that the road to Cane Hill was clear of the enemy, and no contact was expected. It was felt safe enough that the 7th had the horses feeding "with bridles off and girths loosened." The 8th Missouri began to pass by the 7th. Behind these two units, the 1st Arkansas was working its way south to Rhea's Mill to guard the supply trains.

It was at this point that Confederate cavalry came crashing through the Missourians' lines. The Federals were utterly unprepared. They put up a very brief resistance, but it was quickly overwhelmed. "The command...retreated in every direction," reported Captain M. H. Brawner of Company A of the 7th Missouri, "quite a number running into the rebel lines, being killed or captured."

The Missouri Federals fled in complete disorder back up the road to Fayetteville and directly into the First Arkansas Union Cavalry. Panic then seized the Arkansans, and they turned and fled in total disorder. Colonel Harrison wrote afterward that he was abandoned "in the extreme rear of my men who had all left me." [9] The twenty regimental wagons were left behind, and only the one bearing the regimental flag was able to be brought out safely from the chaos.

"The fight grows intensely interesting," reported Confederate General Jo Shelby, commander of the onrushing Rebels, "and my men, feeling the inspiration of the scene, dash on and on, taking prisoners, capturing guns, colors, horses, mules, and every form and variety of clothing, left in the desperate flight of the terror [stricken] enemy."

This was a rough situation. Those who were not fast enough were getting killed or injured or captured. Private Curry Moore of Company K, a member of the 1st Arkansas for a month, was killed. So also were Corporal George Nelson of Company A and Private John Strohan of Company I. Private James Weldon of Company B was so badly wounded in the left leg that he was afterward disabled and discharged. Corporal Hiram Shahan had his right thumb shot off close to the hand. About 45 men of the regiment were captured.

The retreating Federals fled headlong for some four to five miles. They rushed through next regiment, the 1st Missouri Union Cavalry, which tried to stem the tide. Captain Amos L. Burrows reported:

> [W]e saw a large body of the First Arkansas and Seventh Missouri Cavalry on the retreat. We undertook to stop them, and, finding it being of no use, Major Hubbard ordered the fence to be thrown down on the lefthand side of the road, and drew up in line of battle in the wheat-field, and instructed the First Arkansas and Seventh Missouri to form in our rear. They partially did so, but, having several shots fired at our line by the enemy, they broke and fled.[10]

Benjamin F. McIntyre of the 19th Iowa Infantry was with General Herron's main army. As his regiment hurried to the front, he wrote, "the 1st Arkansas Cavalry came rushing by us on their horses completely panic-stricken - many without hats or coats, spurring their animals to the utmost speed. Our force had no influence on them to return and pell mell completely frightened they rushed on by us. This was not encouraging to raw troops and this display on the eve of a fight by Union Cavalry was no such a display of galantry [sic] as we had anticipated and I heard the word Coward fall from more than one boy's lip...."[11]

Finally, the flight reached General Herron. "They came back on me 6 miles south of Fayetteville, at 7 a.m.," he wrote, "closely pursued by at least 3,000 cavalry. It was with the very greatest difficulty that we got them [the retreating Federal cavalry] checked, and prevented a general stampede of the battery horses; but after some hard talking, and my finally shooting one cowardly whelp off his horse, they halted."[12]

The First Arkansas was diverted into a field by the Walnut Grove Church to re-group. "After I had gained the advance portions of my men," Colonel Harrison wrote, "I halted before them with revolver drawn in a number of instances and with the assistance of several of my officers partially stayed the stampede."

This, then, was the first battle for the First Arkansas Union Cavalry. It

Union General Francis J. Herron personally shot one of the men of the First Arkansas Union Cavalry, a "cowardly whelp," as they retreated in disorder at the Battle of Prairie Grove on December 7, 1862. (Library of Congress.)

was an unmitigated fiasco. General Blunt in his after-battle report not only contemptuously chastised Harrison's evening note that his regiment was too tired to move on, but then condemned his getting caught in the Confederate's early morning attack. "Had he, instead of making unnecessary delay, promptly obeyed that order [to move to Rhea's Mill], he would not have had a portion of his command and transportation captured."

The story of the First Arkansas Union Cavalry at the Battle of Prairie Grove was called by Lt. Col. Bishop "a chapter of accidents." Colonel Harrison called it a "counter-march." Several officers of the regiment in a joint letter called it a "retrograde movement." What it really was, was an embarrassing, unmitigated fiasco. Four men were killed, four wounded, and forty-seven captured or missing. Commanding generals believed the colonel to be unreliable. Orders had been perceived to be disobeyed. The regiment had fled in panic. General Blunt in his after-battle report not only contemptuously chastised Harrison's evening note that his regiment was too tired to move on, but then condemned his getting caught in the Confederate's early morning attack. "Had he, instead of making unnecessary delay, promptly obeyed that order [to move to Rhea's Mill], he would not have had a portion of his command and transportation captured."

Perhaps the criticism had been right - that loyal Arkansas men would not, could not, fight.

With no help from the First Arkansas Union Cavalry, the Union Army won the battle of Prairie Grove. The generals thereafter decided to extend their presence in northwest Arkansas by establishing a Federal outpost at Fayetteville for the first time. Notwithstanding his and the regiment's dismal performances, Colonel Harrison was placed in command. Many of the men were from that very town and county, and they were selected to occupy their home ground.

Except for a period of five months, the First Arkansas would remain stationed in Fayetteville from December 1862 until the end of the war. This was a unique assignment for a Union regiment in the Civil War for two reasons. First, it was posted in the same location for more than two years, and second, it was assigned to occupy the very area from which the men had come. These two factors would prove to be both a blessing and a curse for the regiment and for northwest Arkansas.

The poor reputation of the First Arkansas continued. Colonel William A. Phillips was in command of the district, which consisted of his own command on the border of the Indian (Oklahoma) Territory and of Harrison's post in Fayetteville. On March 27th he reported to Major General Samuel R. Curtis, the commander of the Department of the Missouri. "The enemy still hold Clarksville, although no point on this side of the river above it. They hold Fort Smith, from which I could very easily dispossess them, but have been embarrassed by the Arkansas command at Fayetteville.... The report of the Arkansas command, Colonel Harrison, shows several things I have been trying to correct. I believe Colonel Harrison does the best he can with it, and I hope that more rigid discipline may be gradually introduced."

Without any doubt, Harrison did have disciplinary problems with the First Arkansas Union Cavalry. One of the primary causes was that the men were stationed at a post very near their own homes. Soldiers were coming and going to their families without regard to military necessity or permission.

Another factor was that these men were from backwoods hill country where military discipline was about the last thing they cared about. An interesting order was put out by Colonel Harrison to help deal with their near-lawlessness:

> *Head Quarters, Post of Fayetteville Ark.*
> *March 11/63*
> *General Order No. 10*

> *The firing of arms at any hour of the day or night without written permission from a company Commander..., the cutting or injuring of shade trees, shrubbery or fruit trees; the demolishing of fences or the trespassing in any manner upon private property...will be severely punished. All officers and guards are ordered to make it their especial duty to arrest and place in confinement any person or persons found guilty of violating this order....*
>
> *By Order of*
> *Col. M. La Rue Harrison, Commanding Post*

Such was the early story of the First Arkansas Union Cavalry.

Chapter 3
A Cruel Civil War Among Neighbors

On the very day of the battle of Prairie Grove, December 7, 1862, an incident occurred which foretold the terrible turn that the war was taking in northwestern Arkansas. Guilford Center, a 42-year-old private in Company D of the First Arkansas Union Cavalry, had a 17-year-old son named Mark. The family home was close enough to the battlefield that Mark walked over to see it for himself. On the way back home he was stopped by a group of Rebels.

What the men wanted to know was where Mark's father and his regiment were. The boy would not tell (and probably did not know anyway) so his interrogators put a rope around his neck and hanged him. Before he lost consciousness, however, they lowered Mark to the ground and questioned him again. Three times this was done, but the boy would not talk. The tormenters gave up and let him go.

This was but a single page in what would become a lengthy book of despicable outrages. Unionists and Confederates alike not only fell victim to these atrocities, but perpetrated them as well. War is by definition a very mean business; civil war is even meaner. In northwest Arkansas it was not just a "war between the states," but a true civil war among neighbors, and a cruel one at that.

This was severely manifested when the Confederate authorities were directed by General Thomas C. Hindman not to accord Union prisoners from Arkansas the usual rights given Northern prisoners of war. Since nearly fifty men of the First Arkansas Union Cavalry were captured at the battle of Prairie Grove, this was a matter of the utmost importance.

The custom of that time was to "parole" captured enemy soldiers. This consisted of the captured men giving their oaths that they would not take up arms again until they had been "exchanged." The parolee would then be released to return home until a commission of Union and Confederate representatives would agree to an exchange, whereby so many Northern and Southern troops would be released from their paroles. Once exchanged, the "prisoner" would be notified to return to his unit. The system had its difficulties, but in an era when personal honor was sacred, it generally worked.

Most of the captured First Arkansas Union Cavalry members were paroled before word was put out not to do so. A few unfortunate men were then not paroled, and they met a harsh fate. Privates Joshua W. Day and James Echols of Company A, both of Benton County, Arkansas, were taken to Fort Smith and executed.

Corporal Charles B. O'Neal of Company G, from Carroll County, Arkansas, met the same fate on Christmas Eve. Frederick Miles of Company H was taken to Forth Smith and died there on Christmas Day, probably by execution. Anson Hodges of Company D was taken to prison in Little Rock. He escaped the following month, but was killed fifty miles north of there by Confederate home guards.

These cruelties were equaled in return by members of the First Arkansas Union Cavalry. Private William F. Robinson of Company B was a Missouri teenager with friends who had chosen the Southern cause. He did not hold it against them personally, but others were not so broad-minded.

In the days following the battle of Prairie Grove, Confederate soldiers were paroled and permitted to return home to Missouri. Robinson was on escort duty for supply wagons heading south from Springfield when his train encountered such parolees heading north toward home.

"I noticed that when we met these men," he recalled, "and as they passed on, some few of the most desperate and wicked men of our command would drop back. After they had gone on, we could hear the distant report of guns and it was not long before it leaked out that these men were murdering paroled Confederates and that a few of the officers encouraged the dirty and brutal work of shooting these unarmed and defenseless men. Some of the men and officers protested against this barbarity but it did little good."

After this had gone on for a period of time, Robinson encountered two of his neighbors who were paroled Rebels. When the murderers fell back to carry out their evil work, Robinson and his brother Ezekial did the same. They caught four Union soldiers just as they were about to shoot, intervened, and saved the unarmed Confederates.

Such incidents of hate and brutality continued throughout the years of the war. It must be remembered that many of the men of the First Arkansas Union Cavalry had been driven from their homes by Confederates, and they were back for what Lt. Col. Bishop called "the *avenging* strife."

A major factor in the deterioration of the war into this brutal civil strife was the national policy of both the United States and the Confederacy to concentrate their resources and attention on matters east of the Mississippi River. The result of these policies was that neither side put enough regular uniformed soldiers into northwest Arkansas to maintain any kind of military or social order Instead, both sides called upon their respective sympathizers to carry on the war in ungoverned bands of bushwhackers. The Federal troops were insufficient in numbers to truly occupy the area, and only enough to establish a vulnerable outpost at Fayetteville. For their part, the Confederates after December 1862 kept no regular soldiers in northwestern Arkansas except for raids.

In order to maintain any sort of presence whatever in northwest Arkansas, Confederates were forced to rely on largely autonomous guerrilla units. The "partisan rangers" authorized on paper by an order issued on June 17, 1862, were to prove the curse that forever haunted the First Arkansas Union Cavalry.

These bands of irregulars were for the most part on their own, out of control of Confederate authorities, and yet permitted and encouraged to wage warfare on Federal troops. The Union soldiers considered these bands as outlaws rather than soldiers, and their primary assignment was to exterminate them. Men on both sides were often lawless and ruthless, and a genuine hatred grew between them. Each had legitimate complaints of wrongdoing against the other.

The situation on both sides, and of those caught in between, was most difficult. Some civilians were organized into their own spontaneous Confederate "companies" under the authority of General Hindman. Unionist civilians organized themselves into units called "home guards," which were the equivalent of the Rebel bands of irregulars. Other civilians, victimized by both sides, felt compelled to arm themselves for simple protection. Federal soldiers could at any time be attacked by men in civilian dress, however, and so had an inclination to call any armed civilian a "bushwhacker" or "guerrilla."

The Confederate Order
Authorizing Guerrilla Bands in Arkansas
to Operate Against the Union Army

General Orders No. 17
Head Quarters, Trans-Mississippi District,
Little Rock, Ark., June 17, 1862

I. For the more effectual annoyance of the enemy upon our rivers and in our mountains and woods, all citizens of this district, who are not subject to conscription, are called upon to organize themselves into independent companies of mounted men or infantry, as they prefer, arming and equipping themselves, and to serve in that part of the district to which they belong.

II. When as many as ten men come together for this purpose, they may organize by electing a captain, one sergeant and one corporal, and will at once commence operations against the enemy, without waiting for special instructions. Their duty will be to cut off Federal pickets, scouts, foraging parties, and trains, and to kill pilots and others on gun-boats and transports, attacking them day and night, and using the greatest vigor in their movements. As soon as the company attains the strength required by law, it will proceed to elect the other officers to which it is entitled. All such organizations will be reported to these head-quarters as soon as practicable. They will receive pay and allowances for subsistence and forage, for the time actually in the field, as established by affidavits of their captains.

III. These companies will be governed in all respects by the same regulations as other troops.

Captains will be held responsible for the good conduct and efficiency of their men, and will report to these head-quarters from time to time.

By command of Major-General Hindman

[Bishop, Albert W., *Loyalty on the Frontier*, 1863, p. 97.]

The First Arkansas Union Cavalry

On December 21, 1862, Lt. Col. James Stuart of the 10th Illinois Cavalry went from Fayetteville to Huntsville in command of detachments from his own unit, the First Arkansas Union Cavalry and the 8th Missouri Cavalry. Arriving at daybreak on the 22nd, he found that 150 Confederate troops had "committed depredations on all the Union families in that vicinity, more especially that of Judge Murphy, the ladies of whose family they stripped of everything but what was on their bodies, leaving them in a destitute condition."

The situation in the Murphy family was soon brought to a disastrous conclusion. The former Unionist delegate to the secession convention who alone voted "No" had fled to southern Missouri, and had made arrangements for his family to join him. But it turned into the sort of family tragedy that was all too often typical of the war. The hardships of their destitution and exposure to the elements as they fled resulted in daughters Louisa, aged 24, and Laura, 22, dying in St. Louis on March 1st and 13th, 1863. Unionism often carried a very high price in Arkansas.

Families of the First Arkansas Union Cavalry suffered, too. Bushwhackers claiming to be Confederates visited the family farm of Davis Chyle of Company M in Ozark County, Missouri, and took away his 60-year-old father. Forcing him into the woods, they killed him along with others.

On one occasion Private William Bloyed of Company D was home in Washington County, Arkansas, on leave from the regiment. While he was out at the blacksmith forge under a bluff near his home, Confederate partisan rangers came by. This was just the scenario in which Union men could get killed. His quick-witted wife ran into the road yelling, "They went that-away! Hurry and you can catch them!" In their hasty pursuit of imaginary foes, they failed to see their true quarry, Bill Bloyed.

Private Daniel Watkins of Company F was not so lucky. While at his home in Benton County he was killed by guerrillas on March 13th.

South of West Fork, Arkansas, lived part of the extended Caughman family. Five men from the Caughman clan served in the First Arkansas Union Cavalry, and others were apparently in other Federal units as well, so the family's political sentiments were well known. Fourteen-year-old Taylor Caughman was chopping wood at his home when Confederates approached. A conversation of unknown content took place, and then a bushwhacker shot the boy through the heart. The boy's mother and sister, who witnessed the murder, ran screaming to the body as the killers rode away.

Sergeant Jacob Yoes of the First Arkansas Cavalry was in the neighborhood, and with a detachment of men he went after the murderers. They stayed on their trail southward into and through the Boston Mountains and into Crawford County, catching up with them near the town of Dyer.

Sergeant Jacob Yoes. Company D of the First Arkansas Union Cavalry. (Washington County Historical Society.)

Sgt. Yoes found the trigger man and confronted him. The murderer plead for his life, but he received what he had given, a bullet through the heart. As this occurred near the dead man's home, Yoes rode to the home and told the wife that she was now a widow. This was a cruel war, indeed.

Meanwhile, more military-like activities were going on at a brisk pace. Patrols were being sent out, information gathered, bushwhackers tracked down, supply trains escorted, Union families protected and telegraph wires maintained. It was a huge job, and it used up both men and horses. This was shown by the regimental return of the First Arkansas Union Cavalry for December 1862:

> Owing to the great number of enlisted men in the Regt. from all parts of Ark, north and south of the Ark River, and their familiar acquaintance with every road & locality the Regt. has proved of great service to the Army of the Frontier and to the Gen. commanding both as

> *guides and as spies. The Regt. is not perfect in discipline nor has it had the opportunity of other Cav. Regts. Since the 14 of August the 1st Battalion under Major Johnson has been in the saddle almost constantly. The remainder of the Regt has seen no rest since Oct 8/62. The number of serviceable horses in the Regt shows conclusively the amount of scouting performed.*

Fifty men of the First Arkansas under Captain Galloway and Lieutenant Wilhite accompanied General Herron in his advance to Van Buren, Arkansas, and participated in the charge into town there in late December of 1862. Then Company K was attached to General John M. Schofield's escort in January 1863 as he toured northwest Arkansas and south-west Missouri. It returned to Fayetteville on February 5th.

"On the 8th of January," wrote Lt. Col. Bishop later, "a detachment, under command of Lieutenants [Robert N.] Thompson and [John] Vaughan, participated in the defeat of Marmaduke at Springfield, Missouri; Lieutenant Vaughan and Sergeant L. D. Jernigan were severely wounded during the engagement." Nothing is available as to their specific involvement in that battle.

On Sunday, January 25, 1863, the First Arkansas Union Cavalry scored its most successful day of the war in terms of captured Confederates. Colonel Harrison sent out a force from Fayetteville on the 23rd under the command of Lt. Col. James Stuart of the 10th Illinois Cavalry. In his command he had 90 men under Captain Charles Galloway of the First Arkansas, 40 men from the 10th Illinois, and two howitzers.

Arriving at Van Buren on the 24th, the Federals learned that a steamboat had gone to Fort Smith, just a few miles away, for men and supplies. Placing a guard on the Arkansas River bank to watch for its return to Little Rock, the remainder of the troops went into Van Buren, capturing 25 Confederates. The next morning, presumably by the use, or threatened use, of his howitzer and small arms, Stuart brought the hospital steamer *Julia Roane* under his control. He captured 3 lieutenants and 246 men, most of whom were sick. All were paroled to be exchanged for Union troops and the ship was allowed to proceed. Rebels were contemptuous of the Federals capturing a boatload of wounded and sick men.

On January 31st Colonel Harrison sent Captain Charles Galloway of

Company E to Huntsville to protect an anticipated Union rally there. It was held that night and reportedly attended by a thousand people. Galloway was then under orders to proceed to a certain canebrake at Threlkeld's Ferry, surround it and capture the Peter Mankins guerila band. Much to Harrison's distress, however, Galloway did not follow those orders. Instead, he went first to Ozark in pursuit of a steamer with a hundred Confederates, then left the canebrake assignment to Captain Robert E. Travis of Company M, a junior captain, and only seven men.

Both of the revised ventures failed. A Confederate escort prevented Galloway from capturing the ship and Mankins and his men got away. On his assignment Captain Travis was killed along with three privates of the 10th Illinois. Corporal Noel Rutherford of Company D of the First Cavalry was mortally wounded. Travis is buried in the Fayetteville national cemetery.

The grave of Captain Robert Travis at the Fayetteville National Cemetery. (Author's Collection.)

Duty at the Federal outpost at Fayetteville was hazardous. Although there was not a threat of a large battle, individuals could nonetheless be killed at any time. Anyone outside the post in town was always in danger. On December 30, 1862, Private George Hughes of Company B was wounded near Fayetteville. Sergeant William O. Johnson of Company M on February 26, 1863, was in a skirmish at War Eagle Creek. He was hit five times, including once in the knee. His resulting total disability led to his discharge sometime later. On a scout in the Boston Mountains, Private James Hughes of Company B was shot in the head and killed. Private William Patrick of Company E was captured on March 22nd and taken to Little Rock. He was killed while attempting to escape from prison there. Guerillas killed Sergeant Robert B. Kelly of Company I near Cross Hollows, Missouri, on March 30th. On April 9th Private John C. Carrigan was wounded in a skirmish at the middle fork of the White

River, and died the next day.

On February 19, 1863, there was some sort of a mutinous incident at Fayetteville. Lieutenant Robert Mack of Company G was apparently ordered out on a patrol or other detail with detachments from Companies E and G. Many of the men in the ranks refused to do it and "stacked arms" instead. The reasons for this are now lost. They were arrested and court martialed.

The men of Company E who were involved in the mutiny were Corporal John Miles and Privates Abednego Bage, John A. Bage, James Hobbs, Micajah Johnson, Daniel Messenger, George Stockton, John Stockton and Samuel Stockton. Those from Company G were Corporal Elisha Hale, John Harper, J. M. Jones, Joseph Merrill and Alfred Pitts. There may have been others. From the number involved, the event was a significant one.

The charges brought against Private Jones were typical, consisting of one charge and one specification only:

> Charge I. Mutiny.
> Specification I. In this that the said J. M. Jones, a Private of Company G, First Cavalry Regiment Volunteers, did willfully and in direct violation of the Seventh Article of War, stack his arms and positively refused to perform further military duty, this at Fayetteville Arkansas on the 19th day of February 1863.

The witnesses listed on the charging document were Lieutenant Mack, Colonel M. La Rue Harrison, Major James J. Johnson and Regimental Adjutant Denton D. Stark. The mutineers were apparently liberally dealt with and remained in the regiment. Two years later, despite this one refusal to obey orders, Corporal Miles was promoted to sergeant.

The constant wear and tear of hard cavalry service was reflected in the regimental return of February 1863. "The Regt is cept [sic] constantly on duty," it recorded. "Scouts are sent out almost daily.... Discipline is tolerably good, for drilling we have no time the men are always off on duty." This was confirmed by the company return of Company M for the same month: "[T]he company have furnished details for escorts, messengers, mail carriers and been so constantly on duty that but few horses are left in the company, and owing to the scarcity of forage, those now remaining are nearly all unserviceable."[13]

A single sentence on the special regimental muster roll dated April 12, 1863, gave an excellent summary of the First Arkansas Union Cavalry at this time. "This regiment since last muster has been constantly engaged in scouting, guarding forage and other trains and in doing the general duty of an outpost on the frontier."

Chapter 4
Vindication at the Battle of Fayetteville

At the Federal outpost at Fayetteville the First Arkansas Union Cavalry was in an inherently exposed position. It was fifty miles beyond the nearest Federal base at Cassville, Missouri, and could not be supplied by either train or river boat. It had to be served by wagons over a difficult and vulnerable road through enemy territory. The deteriorating condition of the horses rendered it even more isolated, as a mounted reconnaissance could not be maintained in the surrounding countryside. The post could neither be quickly reinforced nor rapidly withdrawn. It was an inviting target. Its greatest defense was the corresponding weakness of the Rebels in northwest Arkansas.

"I hope you will move on Fayetteville," Confederate General William Steele wrote to General Cabell on March 12, 1863. "My information is that there are only about a thousand men there, and no cannon."[14] Upon receiving the further (though erroneous) information that the Federal troops at Fayetteville were preparing for withdrawal, Cabell decided that the opportune moment for a "dash" had indeed arrived. His reasoning was given in his report after the battle:

> Knowing that our good citizens had burdens imposed on them by the Federal troops too grievous to be borne much longer; that it was necessary for me to visit that section of the country, and having been appealed to by citizens, both male and female, to give them assistance, I determined that I would strike there the very first time that I saw the least hope, whether I succeeded in taking the place or not.
> As soon, therefore, as I learned that Phillips was moving around with his Indian brigade to flank General Steele, and, having consulted with General Steele, who agreed with me (and desired that a dash should be made at Fayetteville, if nothing more) that it

> was necessary, and, having heard that they were getting their wagons ready (which proved to be false) to reinforce Phillips, besides being without forage (nothing to feed my horses), I determined to make a bold dash at that den of thieves, and, if possible, to take it.[15]

General Steele later described the attacking force. "General Cabell's brigade," he wrote, "having been assigned to my command, would convey the idea of a respectable force, which is an erroneous idea. Monroe's and Carroll's regiments, both weak, are all that have ever been here. The balance consists of companies and battalions scattered through the country...and...it is a matter of doubt if any great number of them can be brought together."[16]

Brigadier General William L. Cabell led the Confederate forces in the Battle of Fayetteville. (Library of Congress.)

Assembling his brigade at Ozark, about 75 miles by road to the south located along the Arkansas River, Cabell made ready for the assault. Except for the three companies of Colonel Hill's battalion of cavalry which were left behind in Ozark because of unshod horses, the entire brigade was brought along. The Confederate expedition against Fayetteville numbered about 900 men.

The Federal post at Fayetteville was not in a prime condition to receive this Confederate attack. Colonel Harrison had under his command the First Arkansas Union Cavalry, the First Arkansas Union Infantry and the First Arkansas Union Light Artillery. The apparent strength, however, was not actually present. The latter two units were in the process of being formed. In March Colonel William A. Phillips, commander of the Federal District of Northwestern Arkansas, reported, "At the post Fayetteville is the First Arkansas Cavalry, in poor condition; First Arkansas Infantry...and a battery (the men without guns), the latter two forces being of no consequence at present. The whole force is not there; is not as great as it appears on paper..."

Harrison gave his own assessment of the strength of his command on April 1st, two weeks before the battle:

> 1st. The state of my command. The First Arkansas Cavalry numbers an aggregate of 1,032 men; probably when all are at the post they may number 850 effective men. They have 154 serviceable horses and 65 unserviceable, all told. The regiment has not received any clothing for three months, and only a very small supply since November, so that a large part of the men are in a destitute condition.
> The First Arkansas Infantry will number in a few days an aggregate of 830 men; probably 700 of them effective. They are totally without transportation, clothing or tents, or equipments of any kind, except the arms picked up on the Prairie Grove battle-ground, which are of all patterns and calibers. The destitution of clothing is very great, and much suffering and sickness prevails on account of it; besides it would be a ruinous policy to place this undrilled, barefooted, butternut regiment in the field to be mixed up and cut in pieces by rebels in the same dress.
> The First Arkansas Light Artillery numbers 110 men, who are destitute of clothing, and have never received their guns. Of course, nothing can be expected of them...[17]

Rumors were rife that the Rebels would strike at Fayetteville. On April 2nd, Phillips wrote to Harrison ordering him to be prepared: "Call in the command, and keep it at the post. Throw up earthworks as speedily as possible. Defend yourselves as you see fit, but lose no time.... Put your men in effective shape. Make their position strong. Exert yourself so as not to embarrass me.... I...have confidence in your judgment with *your peculiar command.*"[18]

The old Yankee suspicion about Federal Arkansas troops is apparent in this "peculiar command" comment. Loyal Arkansas soldiers still had a point to prove to their fellow countrymen - that they could and would fight. Their opportunity to do so was fast approaching.

At the time of the Battle of Fayetteville, a significant part of the First Arkansas Union Cavalry was not present. Of its 45 officers, twelve were gone. Eight were on detached service on scouts. The other four were absent with leave, absent without leave, "in the hands of civil authorities" (whatever that meant), and under arrest.

Nearly half of Company D was gone on a foot scout under Lieutenant Wilhite in the Boston Mountains. Company F had only 16 men present under Sergeant John Dienst, with the remainder absent with Captain Richard H. Wimpy on escort duty to Cassville, Missouri. Company M had only 13 men present as the rest were under Lieutenant John B. Turman on forage duty. Company H had no officers present and was under the command of 2nd Master Sergeant John G. Black.

Of the cavalry enlisted men, 976 were on the books at the time, but a considerable number of them were away on the day of battle. This included 62 sick, 23 under arrest or confinement, 20 absent with leave, 84 absent without leave, and 223 on detached service. That left 554 men available for any battle.

The First Arkansas Union Infantry was in a similar condition. On the day of the battle the First Arkansas Union Light Artillery was entirely gone to Missouri. All things considered, some 900 Confederates were attacking about 1,116 Federals.[19] In an approximately one-to-one attack, it was obviously going to be very tough going for the Confederates, but, as Cabell stated afterward, he had mistakenly expected the Union troops to be in the act of vacating the town. If this had been true, then the odds would likely have shifted in his favor. But it was not true. Even as it was, however, Cabell still had two things in his favor - surprise and artillery. If he was to win this battle, he would need to make good use of both.

Cabell's Brigade left Ozark at about three in the morning on

Thursday, April 16th, with three days rations and a full supply of ammunition. On Friday they rode till noon, then rested until sunset. In the dark, they made their way toward Fayetteville up the Frog Bayou Road.

Although the Union defenders at Fayetteville were expecting a Confederate attack to occur sometime in the near future, they were not in immediate expectation of it on the evening before it actually occurred. Lieutenant Joseph S. Robb of Company B of the First Union Cavalry returned to camp on Friday the 17th, reporting that his scout in the direction of Ozark revealed "no apparent preparations of the enemy to move in this direction." Colonel Harrison took the report at face value. Lacking the horses to maintain a constant mounted patrol, he decided to wait until the next day to send out another scout. Lt. Robb, by a very considerable margin, had missed the very purpose of his scout, returning to his post with a Confederate army only hours behind him.

Throughout Friday night, April 17th-18th, the Confederates closed in on Fayetteville. Moving along the West Fork of the White River, they advanced northward and then westward toward town. One of the people guiding the Rebels that night was Sergeant Mathew W. Sumner of Company A of the First Arkansas Union Infantry, who had deserted from Fayetteville just two days before and gone over to the Confederates.[20] Sumner, who must have known the most recent Federal dispositions around town, was likely a very great aid to General Cabell.

Moving in closer to Fayetteville, a few minutes after sunrise the Confederates encountered the dismounted Union picket just east of town. It was just about sunrise now, about 5:40 a.m., and it was time for the attackers to make their move.

The Rebels quickly overran the picket post, but shots were fired before it was all over. Privates Jonas Riddle of Company A and Lucien Amos of Company H were killed. Private George W. Russell of Company G was captured, identified as having been a Federal spy in the past, and hanged in execution. One or two others were also captured.

Though killed and captured, the Union guards had nonetheless successfully performed their mission. "The firing of the picket had alarmed the command," Colonel Harrison wrote afterward. General Cabell's element of surprise, so carefully preserved up to this moment, now began rapidly to dissipate.

With the enemy alarmed, the Confederates moved as quickly as possible to get into a position on the south and west side of East Mountain (now called Mount Sequoyah) to launch their attack. General Cabell, the brigade staff, the artillery and the cavalry reserve turned off the road into town and went northward up a mountain road to a fairly flat point overlooking Fayetteville. The men designated for the attack went a little further west, then turned off the road and went northward up a deep ravine to get closer to the Federal positions. Other cavalry went on to the outskirts of town and set up a far left wing on the southeast edge of the city. These movements were not easy or fast. The men had to be brought up and dispositions completed. The ground was uneven. The time involved in this effort dissipated what little advantage of surprise still lingered after the firing of the picket.

"The soldiers were still in bed when the first alarm was given," wrote Sarah Yeater, who was in the house of Pastor Baxter house that morning. The commotion awakened Federal Lieutenant Elizur B. Harrison, the younger brother of the post commander. "About daybreak on the morning of April 18th," he recalled many years later, "I was awakened by an unusual noise, and hastily dressing, I opened the east door of my room [in the Baxter house] and to my consternation saw near the back of the lot a column of Confederate cavalry. It is needless to say that I hurriedly shut the door and made my getaway through the front door down to College Avenue" (then known as the Telegraph Road and Cassville Road).

Once out of the house, Lt. Harrison found his brother, the Colonel, coming away from Headquarters House where he resided. Joining him, they hurried together over to the camp of the First Union Cavalry, which lay north up the Telegraph Road. "We found the men getting into their clothing, gathering arms and ammunition," the Lieutenant said, "while the officers were getting the men into order."

As previously mentioned, the Federals were not expecting an attack on this particular morning. Colonel Harrison was, as Union Dr.

Seymour D. Carpenter described, "partially taken by surprise." Nonetheless, he responded quickly and "had time to get his men in position before the attack was made." Lt. Harrison later said that his brother "at once took command and very quickly the men...were marched to position in the rear of the Tebbetts house and the dense hedge that surrounded it."

Headquarters House and its historic marker. (Author's Collection.)

Harrison set both the First Cavalry and the First Infantry into motion. He did not know how many men the enemy had, what their troop dispositions were, or what their intentions were. All he could do was react, and he did it well, setting up a main defensive line with protective flanks and a reserve. He ordered "the First Sergeants to personally see that their companies were supplied with ammunition." He ordered 25-year-old Lieutenant Colonel Elhanon J. Searle, who had formed his men on its parade ground, to slowly withdraw his exposed regiment on the east end of town toward the cavalry.

The First Union Cavalry, on foot throughout the battle, was ordered into position to receive the attack. The third battalion of the First Cavalry (consisting of only two companies) was placed on the right, under the command of Major Ezra Fitch. The second battalion under Lt. Col. Albert W. Bishop and Major Thomas J. Hunt was put on the left. The center, commanded personally by Colonel Harrison, consisted of four companies of the First Cavalry with three companies (A, F and H) of the First Infantry in immediate reserve. "Fearing that, not being uniformed, they might be mistaken for the enemy," Harrison put the rest of the Union infantry (seven companies) in a distant reserve in a sheltered position to the rear under the command of Lt. Col. Searle. Captain Rowen E. M. Mack of Company G of the Cavalry (mostly men from Carroll County, Arkansas) was ordered "to reconnoiter on the right to prevent a flank movement in that direction." Captain Hugo C. Botefuhr and Company C were placed in ready reserve.[21]

About six o'clock the Confederates made their initial move toward Fayetteville, charging on horseback "with wild and deafening shouts" out of the ravine and toward Federal Headquarters (at the Tebbetts house) and the Baxter house. This "dashing charge," wrote Cabell, "drove the enemy to their pits and to the houses, where they rallied and poured in a dreadful fire with their long-range guns." John M. Harrell, who later served as a Colonel in the brigade but was not present at this battle, wrote after the war, "In the streets Cabell's men met with effectual resistance from the windows, doorways and corners of the houses...."

Henry G. Orr of Parson's Texas Cavalry wrote that the Confederate attack sent "the enemy seeking protection in and behind houses and some in rifle pits." In the Civil War rifle pits were not foxholes in which the soldier would stand, but rather shallow trenches in which he would lie down and fire, with the dirt from the hole piled in front of him for protection. In this battle they were likely more of an annoyance than a serious military factor.

In defending the attack, First Sergeant William M. Burrow of Company E, First Arkansas Union Cavalry, fell badly wounded. "As his comrades were bearing him from the field, he begged them to 'lay him down and go to fighting,'" wrote Lt. Col. Bishop. Burrow died of his wound two weeks later.[22]

As the Confederates neared Federal Headquarters and the Baxter house, and the Federals occupied their rifle pits and houses, the

Rebel cavalry attack halted so the artillery to get into position. Within thirty minutes of the initial attack, Lieutenant William Hughey had his two-gun section of artillery in place and ready for action. The Confederate guns focused on the camp of the First Union Cavalry. Cabell wrote that the artillery "did frightful execution in the enemy's camp, driving them out and completely scattering their cavalry for awhile." Colonel Harrison reported that the enemy guns "opened a sharp fire of canister and shells upon the camp…, doing some damage to tents and horses, but killing no men."[23]

"We were using the large brick smokehouse in the rear of the Van Horn lot as an arsenal," said Lt. Harrison, "and Jim Bell, a private in Company I, was stationed there to notify us if any attempt was made to capture our supplies." One cannon "shot passed over the higher ground and in falling crashed through the body of Jim Bell at the arsenal."

The effect of the intense artillery fire was to cause a considerable fear and demoralization in many of the untried Federal troops. There were no Union cannons to return fire. Lt. Harrison stated that at its beginning the battle seemed destined to end in defeat for the Federals. It was probably about this time that two men from Company A of the First Union Infantry broke under the stress of their first battle. Thirty-year-old Private Francis W. Cannon "run and concealed himself," and 18-year-old Private Gilbert C. Luper "was frightened and run when the rebels made their appearance."

Corporal Thomas Bingham of Company E of the Infantry and Private Cyrus Barber of Company L of the Cavalry also decided that it was time to flee. Colonel Harrison later reported that he had thirty-five men missing, "mostly stampeded toward Cassville during the engagement." He further noted that Lt. Col. Searle and Major Ham of the First Infantry "did good service in keeping their men in position [in reserve] and preventing them from being terrified by the artillery."[24]

But the fear was not confined to the enlisted men. First Lieutenant Crittenden C. Wells, quartermaster of the First Infantry, "ran away disgracefully to Cassville, Missouri." At the commencement of the bombardment, at about seven o'clock, Captain DeWitt C. Hopkins of Company I and First Lieutenant William L. Messenger of Company D, both of the First Cavalry, quickly lost hope under the cannon fire. In their minds they exaggerated the number of guns they faced. They went to Colonel Harrison, saying that six artillery pieces had been planted by the Confederates, and that the enemy was flanking them.

"I replied," Harrison recorded, "'How can we retreat; they on horses and we on foot? Would you wish to be disgraced by a surrender?'" The spooked officers said they would not, they all vowed to "fight it out to the death," and returned to their commands. "And right well did they do their duty," Harrison said afterward. In response to the concern about flanking, Harrison directed Lt. Col. Searle of the First Union Infantry to send two companies (one of which was probably Company K) to the left of Lt. Col. Bishop's position. He also directed Captain Randall Smith and Company A of the Union Infantry to report to Major Fitch on the right.[25]

The demoralizing effect of the artillery, and the wavering it caused in the Union defense, was Confederate high tide at the Battle of Fayetteville.

With the Union lines being hopefully softened by the artillery bombardment, Cabell now moved to test the resolve of the Union troops, finding out for himself whether Arkansas Union troops would fight or run. Monroe's Confederate cavalry dismounted and advanced on foot as infantry through the open fields northeast of town toward the Federal left. Lt. Col. Bishop, the Union commander in that section, called Monroe's assault a "bold advance." Lieutenant Harrison called it "a heated skirmish."

Union Captain William Johnson with Company M were the primary defenders where the attack struck. Johnson had just two months previously replaced the killed Captain Robert Travis, and just six days earlier had been certified by Dr. Amos Caffee, the regimental assistant surgeon, for 60 days leave of absence to go to St. Louis for surgery. Unfortunately, before he could leave town the battle began.

Lt. Col. Bishop reported that Captain Johnson "had his right arm shattered while leading his men forward under a galling fire" in repelling Monroe's dismounted attack. Dr. Seymour Carpenter later wrote that "Capt. Johnson is both an intelligent and meritorious officer and...his wound was received while leading a charge which decided the fortune of the day." This is the likely time that Private Davis Chyle and Corporal Doctor B. Norris, both of Johnson's Company, were wounded also.[26]

Monroe's dismounted attack did not break the Federal left, and it was discontinued.

Although the strategic objective of the Confederates was the capture of Fayetteville, the primary tactical target was the Federal headquarters house in the center of the Union line. There was ongoing shooting and maneuvering, and the headquarters was repeatedly but unsuccessfully charged by dismounted Rebels. These men were almost certainly Thomson's men from Carroll's Cavalry. Bullet holes in interior doors of the house even today mark the ferocity of the struggle there.

Yankee firepower was overwhelming. General Cabell quickly learned to his chagrin that the Union weaponry was vastly superior to his own. The Federals were "well armed with Springfield and Whitney rifles," and "they poured in a dreadful fire with their longer range rifles." These rifles were one-shot muskets with a rifled, or grooved barrel interior, which would put a spin on the musket ball, shooting it further and more accurately than the old smoothbore guns. The Arkadelphia guns used by the Rebels, Cabell lamented, were "no better than shotguns."[27]

The Confederates had difficulty getting close enough to use their shotguns effectively while always themselves within deadly range of enemy rifles. The Rebels were at a serious disadvantage with their inefficient arms, and, in the final analysis, the disadvantage could not be overcome. Cabell claimed that the Federal rifles could shoot as far as his artillery, which seems an exaggeration but the comment illustrated his frustration with the situation.

The Confederates successfully attacked and captured the Baxter house across the street from Federal Headquarters, taking it early in the battle. It became the anchor of the Confederate center. The absent Pastor William Baxter later described what he had later learned of the action at his home:

> *[A] battle raged round my dwelling, and in*

> sight of it men lay dead and dying. My house, being in full range of the enemy's fire, suffered from their cannonade, and I doubtless left none too soon for the safety of my family. The house which I formerly occupied near the College was seized and held for a time by the enemy, and was struck and pierced by more that fifty shot, shell and bullets....[28]

A Confederate hospital was quickly established on a level area of the mountainside the Rebels occupied. The medical team there had plenty of work to do. The Federals, meanwhile, were taking their own losses. During the battle First Cavalry surgeons, Drs. Amos H. Caffee and Jonathan E. Tefft "were very prompt in sending out their ambulance and directing where they should be driven, doing this while the engagement was still in progress."

There were various actions in different places. The Rebels took possession of the southeast part of town, capturing and destroying a Federal supply train of 10-15 wagons and capturing six escort soldiers. Private James J. Hutchinson was captured at the hospital of the First Union Infantry. Harrison ordered Major Fitch on the right to send a company "to drive in the enemy's pickets at the hospital."

Before long, the Federals were advancing in an attempt to re-take the gains of the Rebels. Colonel Harrison reported:

> At 8 a.m. our center had advanced and occupied the house, yard, and outbuildings, and hedges at my headquarters; the right wing had advanced to the arsenal, and the left occupied the open field on the northeast of town, while the enemy had possession of the whole hill-side east, the Davis [Baxter] place, opposite to, and the grove of, headquarters. This grove was formerly occupied by the building of the Arkansas College.[29]

Harrison decided to try to knock out the Confederate artillery, thereby depriving the Rebels of their best advantage. He sent his brother, Lt. Harrison, with an appropriate order for Lt Col. Bishop. Bishop then selected Lt. Robb of Company L to advance with two companies and try to silence the guns by picking off the individual artillerists with rifle fire.

Things were stalling badly for the Confederates. Cabell's opportunity to win this battle was fast dissipating. About 9:00 o'clock, Colonel James C. Monroe's cavalry regiment was ordered mounted for a direct charge at the Union center.

At that moment Union Dr. Seymour D. Carpenter, who was in charge of the Federal hospital, was on the porch of a house on the Cassville Road behind Major Fitch's Federal right wing. "East of the road was a wide wooded ravine, in which, and screened by the timber, the enemy's cavalry formed for the charge," Carpenter recorded. "Suddenly I heard a tremendous yell, then the clatter of the horses, then the toss of their flags, and then they were upon us." The Confederate cavalry attacked uphill on the Old Missouri Road (modern Dickson Street).

Dr. Carpenter went on to describe the event:

> Major Ezra Fitch, who was in immediate command of the Battalion in front of me, I had always regarded as a dull and stupid sort of man. I particularly disliked him, because he always wore a tall black plume on his slouch hat, but he was something like a tortoise. He required coals to be put on his back before he could get up a move. In the present instance he rose grandly to the occasion. As soon as he heard the yell, he rushed up and down in front of his line, brandishing a revolver in one hand, and the objectionable plume hat in the other, with oaths that would have done credit to the "army in Flanders," he admonished his men to stand steady, to reserve their fire until the enemy reached the brow of the hill, and then to "give 'em hell."

Colonel Harrison personally commanded the Federal center where the Confederate attack was aimed, and was in the neighborhood of the Union Cavalry's Company F when Monroe's cavalry attack came. He ordered the company to "fire low, take good aim and be sure to kill a man every time."

Map of the Battle of Fayetteville

(Prepared by the Author and Dan Quintans.)

Battle Map Summary
(Numbered paragraphs correspond to numbers on the map.)

    1. 5:40 a.m. The Federal pickets east of Fayetteville are quickly overrun by Confederates under Brigadier General William L. Cabell. Shots alert the Union defenders commanded by Colonel M. La Rue Harrison that an attack is imminent.

    2. Using the ravine running north-south between East Mountain and the town, Confederates move closer to the Federal positions. Cabell

sets up his field headquarters and a hospital on the hillside, with artillery to his front and reserves to his rear. Other Rebel troops occupy the southeastern part of Fayetteville.

3. Harrison establishes a defensive line consisting mostly of the dismounted First Arkansas Cavalry. He commands the center, and places Lt. Col. Albert W. Bishop on the left and Major Ezra Fitch on the right.

4. Most of the First Arkansas Union Infantry, which had no uniforms, is placed out of harm's way behind high ground to the rear. Three companies of Infantry, however, form part of Harrison's center line.

5. 6:00 a.m. Colonel John Scott directs a Confederate mounted charge of Dorsey's Missouri Squadron (under Major Caleb Dorsey) and Carroll's First Arkansas Cavalry (under Lt. Col. Lee L. Thomson). The Union defenders disperse into houses, behind hedges and walls, and into rifle pits, and with their superior muskets bring the attack to a halt.

6. 6:30 a.m. Lieutenant William M. Hughey's Confederate two-gun battery opens fire from East Mountain. It pours in a heavy fire nearly panicking the Federals, but they stand firm.

7. Harrison sends two companies of the First Arkansas Union Infantry to protect the Federal left flank.

8. Colonel James Monroe's First Arkansas Confederate Cavalry makes a dismounted attack in the open fields on the Union left. It fails to break the Federal line.

9. There is fighting in and around Federal Headquarters and the Baxter house between dismounted men from Carroll's Cavalry and Harrison's Unionists. This is the area where the most casualties are inflicted. The Confederates capture the Baxter house, but the Federals hang on to Headquarters and its grounds.

10. 9:00 a.m. Colonel Monroe leads a cavalry charge up the Old Missouri Road (Dickson Street), but runs into heavy defending musket and pistol fire from Federal defenders on the right and in front. The attack fails and the cavalrymen make a left turn and retreat.

11. Lt. Robb of the First Arkansas Union Cavalry leads two companies to within musket range of the Confederate artillery and pours in a heavy fire. With ammunition exhausted, Lieutenant Hughey withdraws his artillery from the battle.

12. The Confederate attack has failed. Desultory fighting continues on, but the battle is essentially over a little after 9:00 a.m. Cabell considers burning the town of Fayetteville as he pulls out, but decides not to do so because of Confederate families there. By noon, the Rebel army is gone.

Union Dr. Seymour D. Carpenter, who witnessed the Battle of Fayetteville and wrote a description of it afterward. (Author's Collection.)

Dr. Carpenter continued his narrative of the battle:

> The brow of the hill was about forty yards from the line. In a minute the long line of Cavalry appeared, the Major [Fitch] rushed in front, gave the command to fire, and a sheet of flame from five hundred carbines greeted them; dozens of men and horses went down; I could see the line waver, and the men frantically reining their horses, and swerving to the right and left. They were armed with sabres, and if they had pistols they did not use them. All our men had carbines, and revolvers, and in a minute not a Rebel was in sight, save the killed and wounded.... The Major sped the fleeting guests, with fresh volleys, and then he, and his men began giving assistance to the wounded. Not a man on our side had received a scratch. The whole affair was over in five minutes. It was a most thrilling sight....[30]

As the attackers charged toward Major Fitch's Union troops ahead, they passed Federal troops at Federal Headquarters on their right. This put the Confederates into a "galling" crossfire from the front and right, "piling rebel men and horses in heaps." Quartermaster Sergeant S. D. Haley of the Cavalry's Company D "especially distinguished himself in repelling the cavalry charge by Monroe." It was believed that Haley fired the shot that brought down the Confederate color bearer, resulting in the capture of the Rebel battle flag.

Meanwhile, Lieutenant Robb was advancing with two companies from the Federal left to move in close upon the artillery "for the purpose of silencing if possible the enemy's battery." "Their artillerists and guns were out of sight, hidden by the brush," Colonel Harrison later explained, "and I ordered that after the discharge of their artillery my men should aim and fire their rifles about one foot above the blaze of the discharge."

Private Hugh Cook of Company L took this mission to heart, advancing 200 yards beyond his comrades. He shot two Confederates, captured their muskets and horses, and brought them back to Federal lines in the midst of the fighting. Lt. Col. Bishop made note of his actions in his after-battle report.

The fire of Robb's Federal skirmish line was effective. The Confederate battery lost one killed and several wounded, plus two horses killed and two wounded. By contrast, Company L of the First Union Cavalry lost only one casualty, when Sergeant Benjamin K. Graham was slightly wounded. "Two [Harrison said several] well directed volleys accomplished this object," Bishop reported of the attempt to silence the artillery. "At all events, the guns were limbered with great speed and hastily withdrawn to play no further part in the events of the day. I regard Lt. Robb's conduct as exceedingly daring & well timed." The artillery fell silent about 9 a.m., not long after Monroe's failed charge.[31]

General Cabell had a different version of the withdrawal of the artillery. He had wanted to move the artillery closer for more effective use but could not do so because he did not have adequate small arms to protect the guns. The artillery was withdrawn, he said in his report, because the supply of ammunition was exhausted. Credence is given this statement by Cabell's report a week later that his artillery guns still lacked ammunition. The truth of the matter is probably that the Confederates were indeed running out of ammunition just at the point that Federal rifle fire was getting very hot, and there was no longer any point in remaining.

Whatever the reason, the only Rebel advantage in the battle was gone. In fact, all the Confederate momentum was now spent, though the attackers persistently hung on to the Baxter house at the center of their line. Both wings had been partially pushed back from this center point. Fighting continued for nearly an hour, with skirmishing, reconnoitering parties and stragglers.

With the Confederate artillery withdrawn the battle was left in the unequal contention between Federal and Confederate small arms. Cabell complained:

> *I found it impossible, with the arms I had, after my artillery ammunition was exhausted, to dislodge them from the houses and rifle-pits with the kind of arms my command had without losing all my horses and a large number of my men, as it was impossible to get near enough to them to make our aim effective without a great sacrifice of life, much greater than would have been justifiable under the circumstances....*[32]

Faced with the reality that his attack on Fayetteville had failed, Cabell thought about burning the town to the ground:

> *I could have burned a large part of the town but every house was filled with women and children, a great number of whom were the families of officers and soldiers in our service, and I did not deem it advisable to distress them any further, as their sufferings now are very grievous under the Federal rule.*[33]

Instead, Cabell ordered the Confederates to retreat. "After a hard fight of three hours," recorded the return of Company C of Carroll's Regiment, "and finding that the enemy greatly outnumbered us, we retreated."

"Let it suffice to say," Lt. Col. Bishop of the First Union Cavalry wrote, "that at 10 o'clock a.m. it [the Confederate army] was a broken, disordered aggregate of galloping humanity, fleeing...for the Arkansas river." This was an over-statement. Contrary to Bishop's comment about "broken" and "disordered," Cabell reported that, "I withdrew my command in good order." Inasmuch as there was no pursuit by the victors, the Rebels undoubtedly did retreat in fairly good order.

Cabell hoped the Federals would pursue him into the woods so he could attack them outside their houses and rifle-pits, but they did not accommodate him in this regard. Harrison wisely saw no sense in sending Federal soldiers out on foot to chase Confederates on horses. If there was ever a victorious commander who was justified in not pursuing a defeated enemy, it was Harrison. About the best he could do was send out Captain Hopkins with Company I of the Cavalry to reconnoiter and drive in Rebel pickets. By noon the Confederates had given it up altogether and were in full retreat for Ozark.

The Battle of Fayetteville, Arkansas, was over.

Colonel Harrison was exuberant as the Confederates retreated. The evening after the battle he sent out a preliminary report by telegraph to General Samuel R. Curtis:

> *Arkansas is triumphant! The rebels...attacked Fayetteville at daylight this morning, and, after four hours' desperate fighting, they were completely routed, and retreated in disorder toward Ozark.... Our stores are all safe; not a thing burned or taken from us.*
> *Every officer and man in my command was a hero; no one flinched.*[34]

"Every field and line officer," Harrison wrote the following day in his battle report, "and nearly every enlisted man, fought bravely, and I would not wish to be considered as disparaging any one when I can mention only a few of the many heroic men who sustained so nobly the honor of our flag." He went on to mention several officers by name: "Lieutenant-Colonel Bishop and Majors Fitch and Hunt...led their men coolly up in the face of the enemy's fire, and drove them from their position.... Lieutenant Roseman, Post-Adjutant, and Lieutenant Frank Strong, Acting Adjutant, First cavalry, deserve much praise." Harrison also issued a Victory Proclamation to be read at church services on Sunday, the day after the battle.

Federal losses were reported by Colonel Harrison as 4 killed, 26 wounded, 16 prisoners and 35 missing. A thorough study of the records, however, gives somewhat different figures, in part due to wounded men dying and missing men becoming accounted for. More accurate figures are 10 men killed and mortally wounded, 28 wounded but not mortally, and 26 captured (a figure from Cabell). Of the 38 killed and wounded, 26 were from the First Arkansas Union Cavalry and 12 were from the First Arkansas Union Infantry. Cabell did not take any prisoners with him, but rather paroled them all. This made a total of 64 Federal casualties. This represented approximately 6% of the total Federal force.[35]

With the incomplete records kept then and available now, about the best that can be said is that the Confederates suffered about 70 killed and wounded and 54 captured. Since many of the captured were also wounded, it is possible only to roughly estimate a total of 100 casualties. This was about 11% of the entire Confederate force.[36]

Congratulations from Union commanders came pouring in to the justifiably self-satisfied Union soldiers of Arkansas. The fact that Arkansas' own Unionist sons had attained the victory over Arkansas Confederates was immediately recognized. Wiley Britton of the 6th Kansas Cavalry noted in his diary that, "The loyal Arkansas soldiers are represented to have acted with distinguished bravery throughout the contest."

General Herron, who had scathingly criticized the First Arkansas Union Cavalry four months earlier, was just as quick to send his compliments to Colonel Harrison: "I must congratulate you on the success of yesterday. It augers well for the future of Arkansas when her loyal troops have beaten the enemy in their first encounter. Such success should encourage us, and I hope soon to see 10,000 loyal men of Arkansas arrayed on the side of the Union. You have nobly sustained yourselves, and deserve a country's gratitude."[37]

General Curtis, added his respects:

> Tender my thanks to the soldiers of your command for their gallant conduct in the battle of Fayetteville. You have done nobly. Arkansas vindicates her own honor by repulsing the rebel flag with her own brave sons. Send minute reports, naming the most

deserving officers and men.[38]

One of the inevitable consequences of the battle, however, was dealing with those men whose performances were not up to expectations. Harrison's initial telegram notwithstanding, some of the men did flinch. Lt. Col. Bishop moved quickly to discipline some non-commissioned officers. Two days after the battle the following order was issued:

> Headquarters, 1st Ark Cav Vols
> Fayetteville Ark, April 20, 1863
> Special Order No. 48
>
> Corporal John Reed of Squadron [Company] "A" is hereby reduced to the ranks for cowardice in the face of the enemy during the battle at this place on the 18th inst. With few exceptions, the regiment behaved nobly and the commanding officer takes this occasion to say that he could not have been better pleased with the daring valor of the men under a murderous fire.
>
> By Order of A W Bishop
> Lt Col 1st Ark Cav, Commanding

The next day, April 21st, Sergeants Joseph Bell of Company I and Cyrus Barber of Company L were also "reduced to the ranks for cowardice in the face of the enemy on the 18th...." On the 22nd, Corporal Jerome Bays was "busted" to private. He deserted less than four months later. In contrast, Corporal Elias Blevins was promoted to Sergeant in Company A for his bravery in battle. However, it was an academic exercise as he was so badly wounded that he was hospitalized for the following year and then discharged from the service.

The best summary of what the Battle of Fayetteville meant for the Arkansas Federals was stated by Lt. Col. Albert W. Bishop of the First Arkansas Union Cavalry. Although always partisan in whatever he wrote, he was indisputably accurate when he stated:

> *This engagement, though of minor importance as compared with the contests of the Army of the*

> Potomac, or the struggles that have recently culminated in the capitulation at Vicksburg, is not without its significance. It was the first battle of the war in which the loyal men of Arkansas were alone opposed to the organized treason of the State, and gave a very decided reproof to the rebel slander, that the Union men of Arkansas will not fight.[39]

Even more significant than the Unionists claiming that this lesson had been taught was the concession by the Confederates that the lesson had been learned. The *Little Rock True Democrat* on April 29th carried a report of the battle stating, "The feds were mostly renegade Arkansians and desperate men. They fought well..."[40] In his battle report General William L. Cabell himself gave a very great compliment to the Arkansas Federals:

> The enemy all (both infantry and cavalry) fought well, equally as well as any Federal troops I have ever seen. Although it was thought by a great many that they would make but a feeble stand, the reverse, however, was the case, as they resisted every attack made on them, and, as fast driven out of one house, would occupy another and deliver their fire.[41]

The First Arkansas Union Cavalry stood vindicated before the world.

Casualties of the First Arkansas Union Cavalry at the Battle of Fayetteville

Company	Name	Rank	Wound
A	Burrows, Reuben B.	Pvt.	Killed
A	Fears, Josiah	Corporal	Slightly wounded
A	Hayes, John	Private	Severely wounded right arm
A	Jack, James	Private	Severely wounded
A	Kise, Frederick	Sergeant	Wounded slightly
A	Nail, Jesse	Private	Deserted
A	Riddle, Jonas	Private	Killed
B	Hottanhour, Gustavus	Lieutenant	Captured and paroled
B	Rutherford, John	Private	Captured and paroled
B	Scaggs, Hannibal	Private	Captured and paroled
C	Reed, Robert	Private	Captured
C	Wooten, William	Farrier	Slightly wounded
D	Asbill, John	Private	Severely wounded chest
D	Carter, Adam	Private	Captured, paroled, deserted, returned
D	Lewis, Henry C.	Corporal	Slightly wounded
D	Miller, William	Private	Captured
D	Quinton, William J.	Private	Slightly wounded
D	Strickland, Levi	Private	Captured and paroled
D	Temple, Francis M.	Private	Slightly wounded
E	Burrow, William M.	1st Sergeant	Mortally wounded, died April 26
E	Grubb, John A.	Private	Slightly wounded right side
E	Harp, John A.	Private	Wounded
E	Taylor, Jordan	Private	Severely wounded
G	Davis, William F.	Private	Severely wounded head, discharged
G	Morris, George A.	Sergeant	Slightly wounded foot
G	Russell, George W.	Private	Captured, hanged
H	Amos, Lucien	Private	Killed
H	Blevins, Elias	Corporal	Wounded, totally disabled, discharged
H	Davis, George W.	Private	Mortally wounded, died April 25th
H	York, William J.	Private	Wounded severely left foot, discharged
I	Bell, James D.	Private	Killed
I	Gregg, Allen C.	Private	Deserted
I	Manes, Jesse	Private	Deserted
I	Oxford, Jacob	Private	Captured, paroled, deserted, returned
I	Sisemore, George W.	Private	Deserted
I	Steward, Joshua	Private	Killed
L	Graham, Benjamin K.	Sergeant	Slightly wounded
M	Chyle, Davis	Private	Wounded
M	Johnson, William S.	Captain	Wounded right arm, discharged
M	Norris, Doctor B.	Corporal	Slightly wounded in the head
M	Todd, Elijah	Private	Deserted

Chapter 5
Hard Duty in the Saddle

As it turned out, the victory of Arkansas Union troops at the battle of Fayetteville on April 18, 1863, was only sand in a receding tide of broader events. The growing siege of Vicksburg, Mississippi, was drawing off Federal troops from Missouri, and Colonel Harrison, in continual fear of an overwhelming attack, demanded from his superiors that the Federal outpost at Fayetteville be either reinforced or withdrawn. Five days after his victory, he was certain that his post was about to be overrun. His plea for immediate help to the commander of the Department clearly showed the severe stress he was feeling, which was to a considerable extent based on imaginary Confederate abilities:

> Fayetteville, Ark., April 23, 1863
> General Curtis:
>
> Can we be re-enforced, and that immediately? We can never hold this place without artillery and horses. There is no use in disguising the fact. Last night I was positive that Cabell and the Fort Smith Indians had combined to attack me at daylight. My men stood under arms from midnight until after sunrise. Such an attack is brewing, and will come in force in a few days. We have no stores here; we have nothing to eat, cannot get trains, with good luck, till the 28th. Must we starve, and then have all the conscripts surrender to an overwhelming force, that will shoot them as deserters? We haul forage 45 miles, and weaken our command by large escorts. We can make no reconnaissances or scouts for want of horses, and could not protect our rear and flanks in retreat. The enemy are splendidly mounted. The men are brave, and have achieved a splendid victory, but we must have help or fall back. Answer immediately what I shall do. Colonel Phillips is about 90 miles from here, and of no use to us in case of an attack from Fort Smith. I should have to face the enemy's artillery all the way to get there.
>
> M. La Rue Harrison
> Colonel, Commanding Post[42]

There were no Federal troops to send to help Harrison because northwest Arkansas was simply at the bottom of priorities. As he requested, however, Harrison got his answer immediately: withdraw to Missouri. A near-panic among the Union families in the area resulted. They had in recent months come into the open with their loyalty, had encouraged and fraternized with the Federal troops, and now they were to be left behind to a Confederate resurgence. For many it was just too much. They packed what they could take and headed north with the Federals. Confederate General Cabell reported a hundred refugee wagons.

On April 25, 1863, Fayetteville was abandoned. Confederates simply walked in eleven days later, making their headquarters in the very Tebbetts house they could not take by storm in the battle on April 18th.

The march back to Missouri was undoubtedly a heavy blow to regimental morale. After seven months of restoring the Union flag to northwest Arkansas by occupying first Elkhorn Tavern and then Fayetteville, now they were going back to where they had started. Adding insult to injury was the fact that they had come as proud cavalry but were retreating primarily on foot. On the way, at Elm Springs, Arkansas, Farrier John S. Bridges was killed by bush-whackers.

In their new assignment in Missouri, the First Arkansas Union Cavalry was divided between Springfield and Cassville, and settled down to a routine of duties at its new posts. Scouting for information on Confederate activity and escorting military wagon trains, the weeks passed by.

It became a rumor that in view of the increasing number of Federal soldiers formed and yet forming in Arkansas, that a Brigadier General was to be appointed from the state. As senior commander of an

Arkansas regiment, Colonel Harrison made it known that he desired the position. In a letter dated May 4, 1863, Harrison stated that "I claim to be an Arkansian because I espoused her cause in the darkest hour." This was truly said. His statement that he intended "to make that state my home at the close of war" was not a mere self-serving assertion. That is precisely what he later did. Harrison was the chief architect of Arkansas troops being in the Union Army under their own state banner, and he was then and is now entitled to the credit for it. But there was no

George Washington Howry Company E of the First Arkansas Union Cavalry. (Prairie Grove Battlefield State Park.)

Arkansas brigadier general appointed, and Harrison remained on with the regiment as its colonel.

Arkansas regiment, Colonel Harrison made it known that he desired the position. In a letter dated May 4, 1863, Harrison stated that "I claim to be an Arkansian because I espoused her cause in the darkest hour." This was truly said. His statement that he intended "to make that state my home at the close of war" was not a mere self-serving assertion. That is precisely what he later did. Harrison was the chief architect of Arkansas troops being in the Union Army under their own state banner, and he was then and is now entitled to the credit for it. But there was no Arkansas brigadier general appointed, and Harrison remained on with the regiment as its colonel.

In late June the senior enlisted man in the regiment, Sergeant Major Robert Thomson, reported back to Union lines in Missouri. He had been captured by the Confederates nearly four months earlier, and imprisoned in Little Rock. "On the night of the 27th of May," Thomson reported, "I with three others [from other units] made my escape from prison. The next night we crossed the [Arkansas] river on a raft which we constructed for the purpose. After nineteen days travel through the woods and hills, avoiding roads and habitations of men, I arrived at Pilot Knob Mo."

Probably everyone in the First Arkansas Union Cavalry was glad to see him - except perhaps for Thomas Brooks. Brooks had been promoted from private of Company B to take Thomson's place as Sergeant Major. It was all resolved satisfactorily, however, as Thomson was promoted to First Lieutenant in the newly organized First Arkansas Light Artillery, and Brooks stayed on as sergeant major.

Scouts and skirmishes were the constant fare of the regiment in Missouri. On July 4, 1863, there was a fire fight near Cassville, where a portion of the First Arkansas under Captain Jesse Gilstrap was stationed. Sergeant Jesse Norris of Company D was captured and killed by the Confederates. Eight days later, Captain Joseph S. Robb returned from a scout into northwestern Arkansas. He reported that "the woods are full of Confederates," but in a skirmish at Cross Hollow his only loss was one horse.

That summer General John McNeil, the commander of the District of Southwest Missouri, lost his temper over absenteeism in the First Arkansas. On August 6th he sent out two orders specifically to the regiment. First, "all enlisted men of your Regiment absent without leave [are to] be reported as deserters, and, if apprehended, they will be treated as such." The second order stated that "no more furloughs will be approved.... The number of enlisted men of your regiment, absent on furlough, already exceeds five per centum of the whole number present for duty."

Captain John Gardner of the 2nd Kansas Cavalry was also unhappy with the First Arkansas. Arriving in Cassville on September 1, 1863, he needed an escort to proceed further south to deliver Army dispatches. He requested an escort from Captain Gilstrap, who was in command of the post at the time. Although Gardner wanted to move out at once, he did not get his escort for two days. Of the 75 men under Captain John Worthington who were finally provided, "one-third of the escort were drunk, whooping and hallooing when we left Cassville."

The expedition came to a disastrous conclusion. On the Arkansas-Indian Territory line near Maysville, Gardner's column was ambushed, and the escort mostly fled. Gardner fought on with about 25 men who stayed while Captain Worthington unsuccessfully tried to rally those who fled. Gardner and 22 men of the First Arkansas were captured by Rebel guerilla leader Buck Brown and his band.

Private Richard Connor of Company A was killed in that action. In a separate action that same day, Sergeant Jared Wikle of the detached Company K was killed near Fort Smith. On September 7th Private Enos Mills was killed by bushwhackers at the White River. Three days later, Private Hill B. Bar deserted while on a scout and "went to the enemy."

The tide was again turning as the summer of 1863 progressed. The siege of Vicksburg was favorably resolved for the Federals, and with the fall of Port Hudson a few days later the Mississippi River was wholly in Union hands. The Trans-Mississippi Confederacy was not only cut off from Richmond, but essentially forgotten by it as well. Union troops, however, were moving back into northwest Arkansas.

On September 13, 1863, General James G. Blunt, who had scathingly criticized Harrison for his conduct at the battle of Prairie Grove the previous December, now spoke more favorably. "I would also recommend the sending of Colonel Harrison's regiment (First Arkansas Cavalry)," Blunt wrote to Major General James M. Schofield, "to occupy Northwestern Arkansas, with headquarters at Fayetteville. They understand the country thoroughly, and would be of great service in ridding that part of the country of guerillas, of which there are numerous bands in that locality. They could also protect the telegraph line, if it should be reconstructed."

Things moved quickly, and on Tuesday, September 22, 1863, the First Arkansas Union Cavalry re-occupied Fayetteville. It would never again be driven from the town. Some hearts were gladdened at this return of Union soldiers; most were not. Sarah Yeater was not at all pleased. "The colonel returned with the Arkansas regiment," she wrote in 1910, "and notified me that he wanted my house [the Baxter house] for headquarters again. I replied that I would cheerfully leave the house but would expect him as commander of the post to see that I had another comfortable house in a desirable part of town, where I would not be disturbed and that no other family be allowed to move in with me." It was ultimately decided that Mrs. Yeater could stay where she was, that officers of the regiment would be boarded with her to prevent disturbances by others, and that headquarters would be located at the Tebbetts house across the street.

At the time of the re-occupation of Fayetteville, General McNeil set the tone for the war on Confederate guerilas. "Cruelty to the bushwhacker will be mercy to the loyal and peaceful citizen," he wrote on September 15th. Then he issued an order that the war in southwest Missouri and northwest Arkansas was to be vigorously carried to the enemy:

> Headquarters District of S.W. Mo., Springfield, Mo.,
> September 30, 1863
> General Orders No. 30
>
> In order to promote greater efficiency amongst the troops of this District; to facilitate the only business apparently now left to us, the extermination of the guerillas and bushwhackers, and restore quiet order, and the reign of law to the Country, it is ordered:
> I. That no more officers or men are to remain idle in any post at any time, except when called in for inspection and muster, than are actually necessary for the guarding of the supplies and camp equipage at that post. The balance shall be actively employed in the field, hunting down bushwhackers, guerillas and all those enemies of mankind that are now infesting this District.
> II. The vice of lying idle at Posts, or as it is commonly termed, "holding the post," has become chronic, and must be cured. We cannot perform service useful to our Country, or do honor to ourselves without constantly devoting our time and our energies to the pursuit of, the laying [in] wait for, and, when found, the extermination of the enemy.
> III. All person of notoriously disloyal character, when the facts shall be established to the satisfaction of any commissioned officer in command of a party, shall be compelled to leave the State, and a prompt compliance with the order exacted, under penalty of being treated as guerillas; a report of such orders will, in all cases, be made to these Headquarters.
> IV. Returning soldiers from the rebel army, who shall come to any post or party in the field, and voluntarily surrender themselves, shall be sent to these Headquarters under guard, with a report of the examination of the prisoners in each case; but all men found in arms prowling about the country, firing on troops, or raiding the peaceable inhabitants, will be shot whenever and wherever caught.
> V. All officers in the service of the United States, whether Volunteers, State Militia, or Enrolled Militia, will

be strictly responsible for the execution of these orders.

By order of Brigadier General John McNeil
C. G. Lamont, Asst. Adjutant General

These were harsh terms indeed. Any Federal officer could decide that a person or family was "notoriously disloyal" and compel them then and there to abandon their farms and homes and leave the state. No right of appeal or other recourse was available to those made homeless.

October of 1863 brought an increase in Confederate aggressiveness. General Jo Shelby went on a raid through Arkansas and Missouri which aroused endless consternation on the part of the Federals. Other Rebel leaders such as Colonel William Brooks, formerly of the 34th Arkansas Infantry, and guerilla Buck Brown also became active in the Washington County area. Federal garrisons and scouting expeditions throughout the Trans-Mississippi traded all the information they could gather trying to determine what the Confederates were doing and where they were going.

The First Arkansas Union Cavalry was strung out in three places during this time. Colonel Harrison had most of the regiment with him escorting wagon trains from Cassville to Fayetteville, a vital function in maintaining Fayetteville as an advanced Federal outpost. At all costs, supplies had to be kept coming. Fayetteville itself was commanded by Major Thomas J. Hunt with a few companies of men. Far to the north, in central Missouri, Captain DeWitt C. Hopkins and about forty men from the First Arkansas were giving chase to Jo Shelby.

On Sunday, October 11, 1863, at about 11:30 a.m. Confederate Captain Smithson of Brook's regiment brought in a flag of truce, bearing a message to Major Hunt:

To the Commander of the Federal Troops at Fayetteville:

Sir: Having the town of Fayetteville surrounded by a superior force, and to prevent the effusion of blood, I demand the immediate surrender of the place and the troops within the same. Thirty minutes will be given for a reply.

Very respectfully, W.H. Brooks, Colonel, Commanding

Major Hunt answered by stating that "no surrender would me made without a fight." He strengthened his patrols and increased them in number. Lt. James G. Robinson led a patrol of 12 men which ran into 30 or 40 Confederates at Walker's house on the Huntsville road. The Federals charged, and the Rebels retreated half a mile where they formed into line of battle. The outnumbered Unionists declined the invitation. Lt. John Vaughan of Company I was sent out on the Old Missouri Road but came back drunk without reporting in. This action would cause him a court martial, but he would survive it and serve until the end of the war.

*Thomas J. Hunt
of Washington County, Arkansas.
Major in the
First Arkansas Union Cavalry.
(Author's Collection.)*

Hunt formed his men in the Fayetteville public square, in line of battle, awaiting attack. When the assault did not occur, the men slept on their arms at camp. The square was strengthened with wagons and breastworks, with the intention that if an attack came the men would fall back to it as a final defense.

At 3 a.m. on October 12, 1862, the men were awakened to be ready in case of a daylight attack. It did not come, either then or any time that day. On the night of the 12th all wagons, military and civilian alike, were put into the Fayetteville public square for defense. There the men slept at their assigned posts. When morning dawned on the 13th, still no attack came.

At 2 p.m. a council of Federal officers unanimously agreed to move all the property to the public square, erect breastworks, and wait for the reinforcements that had been sent for. Hunt estimated the Confederates that day to number upwards of one thousand, with Hunt himself commanding less than half that number. He estimated that his men would have less than full rations after two more days. "I am now making the best arrangements of the forces here," he wrote to General McNeil, "in view of defending the place to the last."

Later that day a Federal patrol on the Old Missouri Road was driven back by Rebel fire. The defenders in town readied their muskets, but again the attack did not come. The 14th came and went. That night Captain Archibald Freeburn rode into Fayetteville with 150 Federal reenforcements from Cassville. Freeburn reported that Brooks was gone, and the crisis was over.

The next day, October 15th, Confederate partisans Brooks and Brown led an attack on a wagon train commanded by Colonel Harrison. Sergeant George Vice and Private George W. Dillingham of Company M were both killed, and Private M.V. Jones of the same company was mortally wounded. Harrison arrived in Fayetteville with the supplies on October 18th.

On November 13, 1863, Harrison wrote to General Sanborn, "No organized band of rebels is known to be now in Northwestern Arkansas." This was an overly optimistic statement, and it is hard to understand why he would say such a thing. If Harrison meant by this that the war was winding down to a conclusion in northwest Arkansas, the future would soon prove that he was very much mistaken.

On December 16, 1863, Captain Worthington of Company H left Fayetteville with 112 men from the First Arkansas Cavalry and one howitzer with orders to scout Carroll, Marion and Searcy counties. Skirmishing along the way, they suffered only two men slightly wounded. Then on Christmas Day, 1863, Worthington found himself in a very hot fire fight in the area of Richland, Searcy County, Arkansas. He lost four men killed and four more wounded. The fatalities included Privates John Forehand of Company H and Randolph Holmesly of Company A. Private John B. Crawford of Company C was "killed after surrendering to the rebels." Two of the four wounded, First Sergeant Jesse Rose and Private Larkin Hendricks of Company H, were mortally wounded and both died within a month.

Captain Worthington wrote the following in his report:

> *There is no part of Arkansas where the loyal*

sentiment was stronger at the commencement of the war that in these counties. At the first call for volunteers, the men their homes and joined the Federal army, and their families are now a prey to the refugee rebels of Missouri. A post in the vicinity of Marshall Prairie would completely break up this rebel rendezvous, and do more, I respectfully submit, toward restoring peace to that section of Arkansas than anything else that could be done.

The year 1864 began just as 1863 had ended, with long hard scouts involving brief and random deadly skirmishes. The Federal campaign against the Rebel guerillas was pursued to the limit of endurance of both men and horses. The elderly Robert Mecklin in Fayetteville noted in his diary on February 4, 1864, "The business of killing men still goes bravely on. Scarcely a day passes during which we do not hear of one or more bushwhackers being killed or that some Federals have been killed by them."

Hard-riding scouts were continually going out of Fayetteville and returning. Company M had a particularly active writer of company returns during 1864 to chronicle this activity. At the end of February he reported on the first two months of the year:

Jany. 18. Lt. Roseman started on a scout to Searcy Co Ark with twenty of the Company, they were in two engagements...in which the rebels were defeated with loss - the men were out until Feby 24, and participated in several skirmishes and were constantly on scouting duty - over 100 rebels were killed on the trip. [This number of 100 is most unlikely.] Feby 25 Lt Roseman again went out with 33 men remaining till 29th, the company killed four rebels on the trip.
    Various other detachments of the Company have been in scouts other than those named. In an engagement on White River Washington Co Ark Private John K. Johnson was killed. Since last muster the Company has been constantly scouting. Quite a number of horses have been killed, abandoned, etc. - and a number captured. The Company is in a good scouting condition considering the great scarcity of forage.

The First Arkansas Union Cavalry

In the endless round of random death that was a fact of life in the First Arkansas Union Cavalry, Private William Means of Company H was killed on February 7th.

In February of 1864 Major General Samuel R. Curtis went on an extended tour of his military domain, now called the Department of Kansas. In reporting to General-in-Chief Henry W. Halleck of his findings, he stated on the 15th, "Fayetteville is a high place, easily defended, but much exposed, being very remote from other posts. The troops (First Arkansas Cavalry) are much scattered." Five days later Curtis sent a letter to General W. S. Rosecrans, the commander of the Department of Missouri, saying, "On this [north] side of the Arkansas [River] small bands [of Rebels] from 3 to 50 are occasionally found; several such bands are near Fayetteville, Ark., where the First Arkansas Cavalry holds a very loose and scattered command."

On March 13, 1864, Company G got caught in a fire fight near Berryville, Arkansas. Five of that company's men were from the Littrell family, originally of Lauderdale County, Alabama. Samuel Littrell had been only wounded a year and a half earlier at Yokum Creek, but on this day the family luck ran out. Joseph G. Littrell was shot through the right leg below the knee, and Joseph N. Littrell was killed outright in the fighting. Also, Private Jesse W. Freeman was shot through the breast and thigh, the latter hit breaking his leg. The wounds disabled him to the point that he was later discharged. Ten months later William Littrell of Company A died of wounds received in another skirmish in Huntsville.

In the spring of 1864 a major Federal offensive in Louisiana known as the Red River Expedition was launched. In cooperation with it, Union General Steele was ordered to proceed south from Little Rock as General Nathaniel P. Banks moved north through Louisiana. As part of the overall Union movement, Lt. Col. Albert W. Bishop was sent from Fayetteville with 400 men of the First Arkansas Union Cavalry to garrison Fort Smith. The men remained there until May 19th, when the Red River campaign failed and Bishop returned the men to Fayetteville.

Thursday, April 7, 1864, was the deadliest day of the war for the First Arkansas Union Cavalry. The unlikely scenario for this was that nine men were assigned to guard regimental corrals southwest of Fayetteville not far from the old battlefield of Prairie Grove. During the evening about twenty guerilas under one Lyon struck, killing/executing all nine Federal soldiers.

Company A lost Nelson Caughman, George Herridan, George Hixon, John James, Henry Poor and James Shibley. James lived to tell the story, lingering on until April 30th. Company E lost William Boman, Burley McGlothlin and Moses McGuire. Boman had been in the regiment for less than a month and had not even been formally mustered in yet. McGlothlin had enlisted just two months earlier.

This was just the beginning of a deadly summer. Sergeant William Mannon was killed by guerillas on April 4th in Franklin County. On the 17th, Private Joel Stanley of Company H died of wounds inflicted by bushwhackers. Corporal Thomas Crawford and Private Jacob Willis of Company C were ambushed and killed on the 22nd while carrying military mail in Benton County.

On May 21st Private William McLaughlin of Company M was killed by bushwhackers while absent without leave at his home in Franklin County. A week later, Captain Rowen E. Mack of Company G and Sergeant Robert J. Scott were killed in action in Washington County. Mack had been with the company since it was organized nearly two years before, and this was a real loss to the regiment.

The intensity of the scouting was revealed by the April 1864 return of Company M:

> Mch 3rd Lt Roseman took 38 of the Company and proceeded on a five days scout in which three rebels were killed and Pvt John Hunter was wounded. He returned on the 7th. On the 9th Lt Roseman took 48 of the Company and went on a scout returning on the 12th. On the 13th he took 35 of Company on scout with paymaster to Neosho Mo returning on the 18th. On the 19th Lt Roseman with 30 of Company proceeded to escort Stark's battery from Fayetteville to Ft. Smith. He returned the 24th. Mch. 27th the Co. moved from Fayetteville to Clarksville reaching the latter place 31st. Distance traveled during the month 527 miles.
>
> Apr 5 Pvt Jasper P. Grady was wounded in action on little Piney Ark. Apr 27 Lt Turman with 72 of the Company while repairing the wire on Horse Head Creek 14 miles east of Clarksville was attacked from a high bluff by bushwhackers. Privates John T. Luna and William

[illegible] were wounded. Three horses were killed and three wounded. Lt. Turman had his horse shot from under him. The Company has been constantly scouting. Estimated whole distance marched during Mar & Apr 1050 miles.

There was a serious breakdown of military discipline in the regiment on June 5, 1864. Captain George R. King of Company E was ordered to take out a patrol on the road towards Huntsville, but when underway a dozen or so of the men deserted. King had to bring the detachment back into the post for lack of numbers. A court martial board presided over by Major Hunt heard the cases against the deserters.

Private Theodore Moton of Company H plead guilty to desertion, and was sentenced to forfeit one month's pay and to ten days at hard labor. Private Jacob Gardiner of Company B was found guilty of the same offense, but he received a different penalty. He was to forfeit a half month's pay "and to be marched over the parade ground with play card [placard] on his shoulder marked Deserter in large letters once a day for two days in succession." Private Wilson of Company G was sentenced to the same punishment. Privates Joel Rickets of Company E, Isaac T. Gilbert and James M. Evans of Company I, and T. M. Wilson of Company F were convicted and sentenced to ten successive days of carrying the deserter placard.

Demonstrating that they gave individual consideration for alleged wrongs, the court martial acquitted Privates George Wood, Hamilton Morrison and Charles Dunnell of Company I and Willis Owens of Company H. They were returned to duty.

An amazing demonstration of liberality and indulgence was shown when the courts martial were reviewed at higher levels. The punishments for Gardiner and Wilson were confirmed as to loss of pay but the carrying of the deserter placard was "remitted." The convictions for Gilbert, Rickets, Wilson and Evans were overturned on procedural technicalities and the men were ordered returned to duty. The end result was that for a number of men deserting their detachment, very little punishment ensued.

By late June of 1864 Colonel Harrison was having trouble with the conduct of First Lieutenant James Roseman of Company M. He had been stationed away from head-quarters at Clarksville, Arkansas, a situation which seemed to engender in him a feeling of independence. Harrison was troubled when in March, on the way there, Roseman "mistreated and allowed his company to rob Union citizens."

The immediate problem began, however, over a promotion. On May 28th Roseman requested that 43-year-old Corporal Jeremiah P. Packer be promoted to sergeant and returned to duty with Company M as that particular unit had lost three sergeants by promotions. Packer had been on detached service with the regiment's howitzer unit. Harrison was happy to promote Packer but wanted to keep him with the howitzer unit "as he was a good gunner and understood the management of the howitzers & I could not replace him."

When the order reached Lt. Roseman in the field he rejected it. "This order is not considered valid," Roseman wrote on June 15th, "...because the Comdg officer 'Troops at Fayetteville Ark." is not the person to make promotions, but the Comdg officer 1st Ark. Cav. Vols. is." This was a foolish game, indeed. Roseman was clearly not thinking properly. In his report to Brigadier General Thayer, Harrison answered this by saying, "Roseman knows that I am commanding the Regt. as well as the 'Troops.'"

"His conduct since leaving this place," Harrison contended, "has been mutinous & disrespectful and he relies upon his distance from Regtl Hd. Qrs. to screen him from just punishment." Yet despite this, the Colonel again demonstrated his usually generous nature. When he could have preferred charges for a court martial, he simply requested that Roseman "be returned to his command."

It soon became a moot point. Two days after the letter was written to General Thayer, on June 28, 1864, Lieutenant James Roseman was murdered about midnight by Private Joseph Wisdom of his own regiment. Wisdom escaped into the night and joined the Rebels. Roseman's death was a serious loss to the First Arkansas Union Cavalry. His recent mistakes could not diminish the fact that he had been a hard-riding campaigner who was frequently out on scout. He would be missed.

On July 31, 1864, the quarterly report of the First Arkansas Union Cavalry showed 11 companies at Fayetteville with two 12-pounder howitzers, and one company at Van Buren. The total of troops, both present and absent, was 1,122, of whom only 538 were present for duty.

The regimental deaths continued during the summer. Ambushes and fire fights might occur at any time, and though pitched battles did not occur, one could get just as killed in a quick shootout in the bush as he could in a great campaign. On June 24th Private John R. Downey of Company C was captured by guerillas and executed. Two days later Private William Hixon of Company A was killed in action.

Several skirmishes occurred in the area of Fort Smith and Van Buren, where detachments of the First Arkansas were posted and did extensive scouting. On the first of August Private Elijah Parker was killed by guerillas. On August 7th Private Michael Peters of Company M was mortally wounded near Lacy Bayou, near Van Buren. Three days later, Private George Byrd was killed at Beaver Pond in the same county. Carrying the mail to Fort Smith, Private Bloomington Dotson of Company L was ambushed and killed in Crawford County. Back in Fayetteville, Private Benjamin Cooper of Company F was killed by bushwhackers on August 7th.

The First Arkansas was regularly out hunting down the Confederate irregulars. Lt. Co. Bishop took out two expeditions in the last half of August, but the elusive Rebels could not be cornered and exterminated, or even made to stand and fight. It was a frustrating thing never to be able to really get their hands on the enemy.

Colonel Harrison was personally in command of a wagon train escort when it ran into a heavy skirmish with guerillas at a place called Nubbin Ridge. On October 20, 1864, these 170 Union soldiers were confronted by a large number of men under Buck Brown in Benton County. A two hour fire fight resulted in casualties on both sides. Corporal John Rainey and Private David Stockton of Company E and Private William Marshall of Company A were killed. Private Herod Johnson of Company A was wounded. Sergeant Francis M. Kimes and Private Joseph Reed of Company A, and Private Wesley Pennington of Company I were captured.

Rebel leader Buck Brown then began an encirclement of Fayetteville. He was joined by Colonel William Brooks with his troops, and on October 27, 1864, the shooting began. In his report of the incident, Colonel Harrison succinctly described the action:

Fayetteville was attacked this morning by a

> strong force, who posted themselves at sunrise on the almost inaccessible bluffs of East Mountain, about 1,000 yards east of town, and opened a brisk fire on my camp. I immediately ordered Capt. D. C. Hopkins, supported by Capt. E. B. Harrison, to move up the mountain with a line of dismounted skirmishers. When within about 200 yards of the top of the bluff they engaged the enemy, whom, as soon as soon as their exact position was ascertained, I commenced shelling with a 12-pounder mountain howitzer, causing them to move their position several times.
>
> At the same time, Captain Hopkins and Captain Harrison led their men, less than 100 strong, up the mountain the face of a galling fire from 700 rebels, charging the topmost bluff three times, and the third time driving the enemy from their position. We found 12 rebels dead; among them 1 captain and 2 lieutenants, all of whom are now being buried by my men, who hold the top of the mountain top. We lost in the charge only 3 men seriously and 4 slightly wounded, none killed.... About 9 a.m. I saw a thick cloud of dust rising in the southwest, and soon another rebel column [Buck Brown's] was displayed on that side of the place and commenced a vigorous attack, but a few well-directed shells caused them to fall back. The firing ceased about 12.30 p.m., and the enemy retired.

Private Francis D. Wilburn of the regiment's howitzer detachment was one of the wounded mentioned by Colonel Harrison. He died of his wounds.

The Confederates did not leave the area, however, and Fayetteville was still essentially under siege. Rations were significantly cut. By day and by torchlight at night, Captain Hugo Botefuhr directed improvements upon the fortifications. Further attack could come at any time, and the First Arkansas prepared for it daily.

When the regular Confederate cavalry retreated into northwest Arkansas after their unsuccessful raid with Sterling Price through Missouri, Major General James F. Fagan was detached and sent southward by way of Fayetteville. There, on the morning of November 3, 1864, hostilities commenced again. Harrison described the situation:

> Price detached Fagan with 5,200 men and two pieces of artillery, which force was joined on the march by 1,500 men under Brooks and Brown. They attacked my pickets and commenced bombarding the town with all their boasted chivalry, not giving me the least time to remove families (most of their own at that) nor demanding a surrender. The bombardment was kept up with one 6-pounder rifled gun and one 12-pounder field howitzer until nearly sunset.
>
> Three times the order was given to charge the works, but each time the men on coming within range of my rifles shrank from the assault and fled to a safe position. At sunset the retreat of the enemy commenced and was continued during the whole night....at sunrise on the 4th instant only about 600 remained to cover the retreat.... My loss was 9 wounded - 1 mortally, 8 slightly. The strength of my command during the engagement was 958 volunteers and 170 militia; total 1,128....

It is doubtful that the Confederates ever had any serious intentions of taking Fayetteville in this attack, which is sometimes denoted as "the second battle of Fayetteville." Certainly the casualties are not indicative of a strong intent. With their numerical superiority at the time of Fagan's arrival, they could have taken the place had they really wanted it. Unfortunately, there are no reports on the incident from Brown, Brooks or Fagan, and one can only speculate at what was really going on. Fagan was most likely just making a demonstration against Fayetteville while working his way south from the Missouri campaign. With a Union army in pursuit, the town would only have to be defended if taken. Fagan was gone on the morning of the 4th, and the following day Federal reinforcements arrived and the siege was over.

The year of 1864 ended essentially as it had begun. Twelve hard months had not materially advanced the cause of either side, and for the Federals there was something of a deterioration in the overall position in northwest Arkansas. The year was well summed up from a command perspective by Colonel Harrison, and from a soldier's viewpoint by Private W. C. Peerson of Company B.

Harrison wrote on November 10, 1864:

> The duties devolving upon my command (eleven

> companies of cavalry), which was the only one in a country 110 miles broad and 250 miles long, have been so arduous that with from 100 to 300 horses (the greatest number at any one time on hand during the summer and autumn) it has been impossible to carry mails..., keep the telegraph in repair, forage for the post, escort supply trains, and at the same time to do the amount of scouting necessary to keep the country rid of the roving bands of the enemy.

Private Peerson wrote a brief description of his service as an ordinary soldier in the First Arkansas Union Cavalry on January 12, 1865:

> We have had a very hard time since we have been stationed at this Post [Fayetteville]. Had a great deal of hard fighting to do. We have been surrounded for days that we darsen to show ourselves outside the Post. I have been in more than a dozen hard fights with the Rebels since I have been in service. Have had two horses shot from under me by the Rebels though I have had the good luck to not get wounded myself, although I have had several balls shot through my clothes. I have at this time 9 months to serve before my time is out and I sincerely hope that I may have as good luck for the rest of my time.... And if I should live through and get my discharge, I shall feel that I have done my part for my country. If every man in the Army had done as much as I have already done, this cruel war would be over.

Chapter 6
Life in the
First Arkansas Union Cavalry

The life of the officers and men of the First Arkansas Union Cavalry was in most ways the same as that of a Northern regiment from, say, Minnesota or New Hampshire. However, in two important respects the stress was considerably greater for those Federal troops from Arkansas.

First, a Minnesota soldier knew his cause was just because he not only felt so personally, but also because his neighbors told him it was. With fairly minor dissent, he marched away to war with the sure conviction that he was doing what was truly right, and with the strong support of his family and friends. Not so for the Arkansas bluecoat. His extended family, and perhaps even his immediate family, as well as his friends and neighbors, were likely to be supporting the Confederacy. While professing patriotism for his nation, the Arkansas Union soldier was considered a traitor by most of his community. Though not a problem for some Arkansas soldiers, others were severely torn between two causes.

Regimental records of individual soldiers disclose a host of fascinating facts about the men of the First Arksnsas Union Cavalry. Private James Cavin of Company B well illustrates the internal personal conflict. He enlisted in the First Arkansas Union Cavalry from Washington County, Arkansas, on September 25, 1862. While on a scouting expedition on November 9th, he deserted, taking with him his valuable Whitney rifle. The regimental return showed that he "joined the rebel army." The following month he fought against the Union forces at the Battle of Prairie Grove.

A year and a half later, in the spring of 1864, a man showed up in the camp of the First Arkansas with a note. The prodigal soldier James Cavin had returned, sending a messenger with a note to ask permission to rejoin the regiment. The messenger shared the note with Colonel Harrison:

...I want you to go and sea Cornal Harrison for me and talk to him and git him to write me a few lines what

> he will do[.] I want to git in the federal army and I am willing to go in enny where...and I will stay and make a good solger as long as I live or the war last.... I want you to tolk to the Cornal[.] if he wont let me come back if he is willing for me to inlist in some other ridgement.... I do not want to give up to be kild.... [sic]

Colonel Harrison made no promises to Cavin, but told him if he would come in and show penitence by his actions, executive clemency would be recommended. Cavin took his chances and came into camp. However, his fellow Arkansans in the company did not want him, and Cavin was sent to the Company H of the 9th Kansas Cavalry.

When General John M. Thayer, then commander of the District of the Frontier, wanted charges preferred, Captain Botefuhr of Cavin's original company complied. Harrison, demonstrating a charity and forbearance which he frequently displayed in the course of the war, wrote to the General. "I would respectfully request that if possible he be restored to duty.... I believe ignorance is his principal fault."

The matter of Private Cavin was never concluded. The war ended with him still serving in the Kansas regiment. Yet his case well demonstrates the tug on the loyalties of some of the men of the First Arkansas Union Cavalry.

The second difference between the Arkansas Federal soldier and his Northern compatriot was that the New Hampshire soldier knew his family and loved ones back home were surrounded by supportive neighbors. When he left for the war, he was firm in the knowledge that his family was safe. But not so for the Arkansas Unionist. His family was subject to reprisal not merely by the social rejection of Confederates, but by the persecution and cruelty of bushwhackers and mean-spirited neighbors. Many men suffered from the well-founded worry over what was happening to their parents, wives and children at home while they were bearing arms against their own state.

Private William Lowe of Company K provides an illustration of this facet of life in the First Arkansas Union Cavalry. In 1863 he submitted the following letter to his company commander:

> Camp 1st Ark Cav, Springfield Mo Sept 4th 1863
> Capt Albert Pearson, Comndg Co K 1st Ark Cav

> Capt:
>
> I beg leave to make you the following statement:
>
> Upon entering the service of the U.S. in your Company, I left at home in Ark. my family consisting of my wife and eight children the oldest of which was fourteen years of age.
>
> My family were driven from home by Geurrillas [sic], causing the death of my wife from exposure. My children are now at this place, without a protector, or any means of support, three of them are sick.
>
> I have spent the time you granted me by furlough, in trying to find them a home without success, and would ask, if consistent with military law and existing orders, that I may be discharged from service, believing that I can be of more service to society at large by taking care of a helpless family than I can as a soldier.
>
> Yours respectfully,
> William Lowe

Captain Pearson endorsed the letter at the bottom: "I certify that the above statement is correct and that the circumstances are as set forth, and would ask if possible that the man be discharged." Colonel Harrison was of a similar opinion. He endorsed on the letter: "This is a case of great distress and for this reason is approved and Respectfully referred to Brig. Gen. J. McNeil Com. Dist. SW Mo." McNeil also approved it.

The day-to-day life in the camp of the First Arkansas Union Cavalry was remarkably fluid. A great many men came and went. This being a war, men were killed and wounded. Also, this being a 19th century army, many men died from, or were discharged due to, a host of various diseases. And the Arkansas men deserted and were absent without leave in epidemic proportions. All of these causes created numerous vacancies in the regiment. Officers and men alike rose in rank, and there was a willingness at the top, when appropriate, to reduce sergeants to the ranks and to dismiss officers from the service of the United States government.

Officers of the Civil War were for the most part a very class-conscious society, possessing what we would now consider an exaggerated code of personal honor. While there were of course exceptions, they often had over-inflated egos in combination with hyper-sensitivity to affronts, real or imagined. Some had a tendency to mind their neighbor's business, making charges of misconduct and causing counter-charges in return. Through the course of the war, the First Arkansas Union Cavalry experienced its share of this sort of activity.

Charles Madison Jones of Company B of the First Arkansas Union Cavalry. (Prairie Grove Battlefield State Park.)

It involved the man at the top of the regiment, Colonel M. La Rue Harrison. Shortly after the battle of Fayetteville on April 28, 1863, Captain Randall Smith of Company A of the First Arkansas Union Infantry, who was slightly wounded in the head in the fighting, formally accused Colonel Harrison of cowardice and dereliction of duty. He cited instances of claimed confusion and incompetence at the battles of Prairie Grove and Fayetteville. He "became so much frightened as to be unable and did not give orders to his men until the action had ceased." In his defense, Harrison set forth a detailed accounting of his behavior on both of those occasions. Officers from both the Cavalry and the Infantry regiments were offered as witnesses in defense of the charges. Although Harrison demanded a court martial to clear his name, this superiors were never concerned enough about the charges to formally look into them.

The matter came to an end when Captain Smith self-destructed. In Missouri in July of 1863 Smith was accused of "conduct unbecoming an officer and gentleman. He visited a private house in the city last night and conducted himself in a very disgraceful manner...." Smith was drunk and disorderly in the presence of enlisted men while on duty as the brigade officer of the day, and was dishonorably dismissed from the Army. There was no further talk of court martialing Harrison. In fact, at the end of the war he was given the brevet rank of Brigadier General.

Like other regiments, a large number of officers were discharged. Some were released at their own request and some were removed by court martial. Captain Bracken Lewis of Company B, a native of Washington County, Arkansas, became one of those who asked to be relieved of command:

> *Head Quarters Co B., 1st Ark Cav*
> *Fayetteville Ark Jan 25th 1863*
> *Col. M. La Rue Harrison*
> *Commanding 1st Ark Cavalry*
>
> *Colonel:*
>
> *I hereby tender my immediate and unconditional resignation as Captain of Co. B. 1st Ark. Cavalry Volunteers.*
> *Cause. Having for a long time previous to my entry into Service been on terms of familiarity with the men of my company I do not feel competent to maintain proper discipline among them, and for the good of the service think some other person should fill my place....*
>
> *Bracken Lewis*
> *Capt Co. B. 1st Ark. Cavalry*

Harrison sent the resignation on to General Schofield, who sent it back because the specific details were not stated. So Harrison frankly supplied those reasons and why he wanted Lewis out of the service. At the same time, he again demonstrated a charitable disposition by asking that Lewis be permitted to go home pending resolution of the resignation. "Captain Lewis maintains no discipline and his Co. has always been in a demoralized state;" Harrison responded, "his incompetency is palpable and he owns it." He went on to say that Lewis allowed his men to go out of camp without leave and did this himself for several days. Captain Bracken was gone by the end of March 1863.

Another officer who recognized his own unfitness for further command was Captain Theodore Youngblood of Company K. On February 9, 1863, he requested that he be discharged "for want of strength & health to undergo the exposure and fatigue of camp life." Colonel Harrison endorsed the letter with the statement that at 49 years of age Youngblood was "too old for the exposure of the field; and will never make an efficient officer." The discharge was granted.

Some charges of military violations were not of such a serious nature that discharge was necessary. Such was the case of Captain Thomas J. Hunt of Company B. Colonel Harrison referred him on December 31, 1862, to General Herron for "passing men of my command to their homes without proper authority." This was a serious problem because the regiment served in the very region from which the men came, and they wanted to be with, and check on the safety of, their families.

Captain Hunt, however, was honest and humble in his response on January 9, 1863:

> General: Relative to the charge against me for passing men through the lines. I did pass two men.... I have never received any orders against such, although it is not proper, it has been a general thing in the Reg't for even line Officers to pass their men ever since I came to it, an error practiced general becomes to [sic] common. All I can say General is that I did it...and am sorry for it and hope you will forgive me.
>
> Yours very truly,
> Th. Hunt

The matter was dropped at that point. Captain Hunt was not only forgiven but promoted to Major eight months later. In April 1865, just as the war was ending, he was further promoted to Lieutenant Colonel when Albert W. Bishop became Adjutant General of Arkansas.

Serious charges were brought against Second Lieutenant Mathew W. Hilbert of Company G. The first charge was "using mutinous and disrespectful language." The specification alleged that "when told that it was his time to go on duty as officer in charge of the Forage Escorts [he] did say - that he would not go on Forage Escort - that he was going on Escort duty to Fort Smith - that he would telegraph higher authority and would show that he had influence at Head Quarters.... This in the presence of enlisted men...on the evening of December 17th, 1863." The second charge was "disobedience to orders" over the same incident. The specification said that Hilbert "did refuse the detail saying that he would go in arrest first, much to the prejudice of good order and discipline."

Lieutenant Hilbert's response to this was to resign by a letter to Major Ezra Fitch, who was in command of the regiment in the temporary absence of Colonel Harrison. "My reasons," Hilbert said, "are that I have unguardedly placed myself in such a situation as has demonstrated my want of self control and feel that justice to the Reg't demands that I should retire from the service. A natural pride makes me anxious to avoid the odium of a Court Marshall [sic]...."

Colonel Harrison evidently had a liking for the Second Lieutenant. Although Major Fitch endorsed the resignation, Harrison rejected it on December 30th. He sent it further up the chain of command "with the recommendation that as this is the first offense, Lt. Hilbert be publicly reprimanded on dress parade and returned to duty. He will also lose his opportunity for promotion to fill the present vacancy in his Company. If in future he conducts himself properly, the past will be overlooked."

Lieutenant Hilbert, however, did not want to be that favorably considered. Either his pride prevented him from getting on with his life, or he just simply wanted out of the Army and was going to take advantage of the situation. In a letter on January 25, 1864, he again submitted his resignation, this time citing that he was passed over and a sergeant from another company was promoted to first lieutenant. Lt. Col. Bishop, commanding in the temporary absence of Harrison, approved it and sent it on to General Sanborn. Sanborn approved it "for the good of the service."

Lieutenant Alfred Hutchison found himself in trouble when he tried to help Corporal John DePriest. DePriest in December 1862 was reduced to the ranks (i.e., to private) for "neglect of duty," and he turned to Hutchison for help. Dr. Henry J. Maynard, the regimental surgeon of the First Arkansas Union Cavalry throughout the war, made the following complaint:

> Fayetteville, Arkansas, 23 January 1863
> M La Rue Harrison, Col Comdg 1st Ark Cav
>
> Sir,

> Last night 2d Lt Alfred M. Hutchison approached me and offered me twenty-five dollars if I would give John DePriest Private Co F 1st Ark Cav a discharge proffering to pay this amount on my promise to make said discharge. On my declining to have anything to do in the matter Lt Hutchison told me that "DePriest had offered him - Hutchison - ninety dollars to procure him a discharge and if you will make it out I will give you half of it." (I give you his exact words.) This offer was repeated.
>
> The above is strictly true and I submit the matter to you that such action may be taken as justice and the good of the service may require.
>
> Respectfully Your Obedt Servt
> H Y Maynard, Surgeon 1st Ark Cav

Not surprisingly, it was Lt. Hutchison who got the discharge. He was held in arrest, released to fight with his regiment when Fayetteville was attacked on April 18, 1863, found guilty of "conduct unbecoming an officer and a gentleman," and dishonorably discharged from the service of the United States. Ironically, it appears from the regimental returns that no action was taken against Private DePriest.

Colonel Harrison himself drew up the charges against Second Lieutenant Jacob Keiser of Company D. The first charge was "conduct unbecoming an officer and a gentleman," the specification of which was that he "did black his face and dress himself in womens clothes and did proceed, thus disguised, to the store of one W.L. Wilson, a merchant of Fayetteville, Arkansas, and did enter said store in search of goods." The second charge of "attempting larceny" was that Keiser "did proceed at the head of a squad of enlisted men...and did forcibly enter the store...in search of goods." This little adventure occurred on Christmas Eve of 1863. It was not long before Lieutenant Keiser was dismissed from the service of the United States.

First Lieutenant Elam O. Kincaid of Company E was arrested on October 26, 1863, and brought before a court martial. The details are sparse in the record, but the charges brought against him were mutinous and disrespectful language, inciting mutiny, and "conduct unbecoming." He was found guilty and recommended for dishonorable discharge, but Kincaid petitioned that the order of dismissal be revoked and that he be returned to duty. General Sanborn agreed to this request, imposing instead a loss of pay for ninety days.

The two most unhappy discharges of all, perhaps, were those of First Lieutenant Thomas Wilhite and Captain Jesse M. Gilstrap. Both were early Arkansas enlistees in the Federal service, both successful recruiters for the regiment, and both known for their personal courage. Yet both were ordered dishonorably dismissed from the service of the United States.

Lt. Wilhite was clearly a favorite of Lt. Col. Bishop, who devoted a chapter to his life in his 1863 book, *Loyalty on the Frontier*. His Unionist defiance in the face of "secession fever" earned the admiration of many. The record is scant as to the offending conduct, but it involved some manifestation of "incompetency," which was given as the cause of his discharge.

Wilhite's friends did not forget him in his hour of need. While his dishonorable discharge was pending, a number of officers got together and petitioned Major General James M. Schofield, the commander of the Department of Missouri, for a reprieve:

> *We the undersigned officers of the 1st Ark Cav Vols feeling an interest in the loyalty of the citizen soldiers of Arkansas...earnestly ask that the case of Lt Wilhite lately ordered to be mustered out...be reconsidered...and...the order...so revoked as to permit him to resign thereby passing him out of the service without personal dishonor.*
>
> *Lt Wilhite was one of the first men of Arkansas who took a positive and determined stand for the Union.*
>
> *He was active and zealous in his efforts to fill the 1st Ark Cav and but few men have contributed more towards the raising of the Regiment than himself. While out recruiting in the summer of 1862 he was cut off in the Boston Mountains and for months was unable to return to the Regiment. A Lieutenancy was tendered him for his meritorious conduct, and serving in the Regiment in that capacity he has always, so far as we know, been obedient to orders.*
>
> *He does not claim to be an educated officer, is distrustful of his executive ability as Commander of a Company, and has already once waived his claim to that*

> *position...in favor of an officer of another company....*
> *He is now absent in comd of scouting party and has no knowledge whatever of the preferring of this request.*

The letter was signed by nine officers of the Regiment, and their petition did not fall upon deaf ears. The pending order of dishonorable discharge of Thomas Wilhite was modified so as to reflect an honorable resignation.

Captain Jesse Gilstrap was a similar case. He was one of the early enlistees from Arkansas, and upon the organization of the regiment was installed as Captain of Company D in July 1862. After a year and a half of service he was put out of the Army for "failing to make proper company returns since his appointment," "lax discipline permitting his men to be disrespectful to him," "sleeping out of his quarters without leave," and "uncleanliness of person to a degree totally unbecoming his position."

Gilstrap petitioned that he be permitted to be honorably discharged. In a detailed letter to Major General Rosecrans, he reviewed his military service and contribution to the cause of the Union. He admitted his failure to make proper returns in a timely fashion, but denied all other accusations. He said the uncleanliness charge "is frivolous and made only to render me contemptible at Head Quarters." The letter was signed by Major Hunt, six captains and several lieutenants of the Regiment along with the surgeon and assistant surgeon. Notably absent was the signature of Colonel Harrison. As with Wilhite, the petition for an honorable release was granted.

Promotions were very liberally given in the First Arkansas Cavalry. There was nothing like a caste system whereby "once an enlisted man, always and enlisted man." Any man of suitable military ability and performance was rewarded as opportunity arose to do so.

James F. Allison of Benton County, Arkansas, enlisted at age 34 as a private in Company F in December 1862. He was promoted to sergeant in April 1863, then that summer to First Sergeant. He was transferred to Company D and promoted to First Lieutenant in February 1864, and a year later to Captain.

Lawson D. Jernigan had a similar journey. He enlisted in July 1862 as a private, and in September was advanced to the rank of sergeant. He was wounded in action January 8, 1863, and was promoted to Second Lieutenant just before the Battle of Fayetteville. He was wounded again on Christmas Day 1863, was promoted to First Lieutenant the following month. A year later, in January 1865, he became a Captain.

Warren W. Munday enlisted at the age of 20 in August 1862. By October he was a sergeant, and in February 1864 was Sergeant Major. The following July he was advanced to Second Lieutenant and the following January to First Lieutenant.

John B. Turman of Company D went from sergeant in 1862 to Second Lieutenant in 1863 to Captain in 1865. Francis M. Ward went from Quartermaster Sergeant of Company I to Captain in a single move. Orin Whitcomb of Company B was promoted from Chief Trumpeter to Second Lieutenant in July 1864. Willis Maynard was promoted from hospital steward to Second Lieutenant of Company L.

Enlisted men who were disciplined were "busted" to private, otherwise known as "reduced to the ranks." Cyrus Barber of Company L enlisted as a private, was promoted to sergeant, then reduced to trumpeter after the Battle of Fayetteville "for cowardice." Later he was stripped of that position "for incompetency." He was just one of a host of men who were promoted to corporal or sergeant and then reduced again to private.

Being "busted" was not always a permanent bar to career advancement, however. Alvin Norris, 22, enlisted in July 1862, and was immediately advanced to corporal. In December he was reduced to private and transferred to another company. There he was promoted the very next month to sergeant when another man of that rank was reduced. In September 1863 he was made First Sergeant, and in March 1865 he became a First Lieutenant.

Another example of a man overcoming a reduction was Joseph Rainey. He enlisted as a private in July 1862, was promoted to First Sergeant in September, then reduced to buck sergeant a year later. He was permitted to recruit for another unit, and in April 1864 he was discharged in order to become a Second Lieutenant in the 2nd Arkansas Union Cavalry.

A considerable percentage of the enlisted men were at one time or another absent without leave (AWOL). This was true of most Civil War

Inspections and Parade Review

Maj. T. J. Hunt, Regtl Inspr, 1st Ark Cav: I call your attention to the following which must [be] strictly adhered to upon inspection.

1st Tents must be in straight files and neatly staked and ditched.
2nd Inside of tents must be perfectly clean, blankets folded, guns and sabres hung on neat racks.
3rd Officers' tents must be neat and books and papers in order, ready to be exhibited.
4th Streets must be thoroughly policed and all offal removed from them as well as from kitchens and quarters.
5th Kitchens must be removed to front (east end) of the tent files and kept neat and in order.
6th Picket ropes must be tightly stretched and horses well tied and groomed and no unserviceable horses at the line.
7th Men must be quiet, orderly and respectful, in and out of quarters, and no signs of waste or negligence seen.

On Parade & Review

1st Horses thoroughly groomed.
2nd Saddles and bridles well arranged without halters & extra straps or ornaments.
3rd Saddle blanket neatly folded.
4th Regulation hats well brushed, no citizens hats or caps.
5th Jackets clean, buttoned up and no buttons lacking.
6th Pants clean and fitting the waist.
7th Boots blacked and spurs bright.
8th Belts clean. Sabres & scabbards & guns thoroughly scoured. Every man completely armed.
9th Every non-commissioned officer with chevrons and stripes.
10th Uniforms complete, with no superfluous ornaments.
11th Blankets and overcoats properly and neatly folded.
12th Accoutrements clean and well fitted.
13th Instruct officer to pay the closest attention to the appearance and deportment of their men and to the policing and arranging of quarters.

regiments, but was even more of a problem with the First Arkansas Cavalry. This was due to the fact that in Fayetteville the men were so close to their homes.

Colonel William A. Phillips, who commanded the district in which Fayetteville was located, wrote to Major General Samuel R. Curtis about the First Arkansas on March 9, 1863. "I visited and reviewed the Arkansas troops three days ago," he wrote. "I was, in the main, pleased with their appearance, but the disposition to go home is too general, and I found it necessary to check it. This has given me a good deal of trouble in the Indian command, but I find the Arkansas command worse than they are."

The usual punishment for being absent without leave was loss of pay and extra duty. Going home without leave, however, was risky in more than just discipline. Private William H. McLaughlin joined the regiment in Fayetteville at the age of 20 in February 1864. He was assigned to Company M, but was AWOL on a visit to his home in Franklin County on May 21st when he was caught by Confederates and killed.

Outright desertion was also a problem. Private John Millsap enlisted at Elkhorn Tavern on November 14, 1862, at the age of 31, and was assigned to Company A. He seems never to have reconciled himself to army life. He deserted in April 1863, was brought back under arrest, and deserted again in April 1864. "This man is supposed to be lurking around his home," one officer wrote in the records, "but I have not been able to catch him." He was later caught and sent to prison.

Private John Davis of Company B deserted on May 26, 1865, by escaping from the stockade while awaiting trial for murdering a black man. From the regimental returns it appears that he was never heard from again. There were many desertions that occurred on a regular basis throughout the war.

Second Lieutenant Cyrus Wills of Company L was sent with Lt. Roseman to get horses in Rolla, Missouri, in June of 1863. Roseman reported back that Wills went to St. Louis in civilian clothes without leave to do so, and never returned.

Private John Miller deserted in February 1863, but was brought back. He deserted again in July 1864, being again brought back in January 1865. He was sentenced to one year's loss of pay and one year's hard labor at a military prison. This was the standard "heavy sentence" for hard-core deserters, and several men suffered the fate. No one in the regiment was executed for desertion, or for any other cause.

In late 1862 the First Arkansas Cavalry began looking for a chaplain. Probably through Lt. Col. Bishop they became acquainted with Private Reuben North of the 2nd Wisconsin Cavalry. North was a minister in the Primitive Methodist Church who had with him a letter of recommendation dated March 3, 1862. In it a number of pastors from his home area stated that "we do cordially commend [him] to Christian people wherever he may go as an earnest and able and laborious preacher of the Gospel - a consistent Christian and an indefatigable and successful laborer in the vineyard of our Lord and Savior Jesus Christ."

The regiment was able to arrange for the transfer of North to the First Arkansas, and he served from February 1863 until mustering out in August 1865 as the regimental chaplain. Records show that he performed a number of weddings, including those of Private Francis M. Temple of Company E on March 29, 1863, Private Francis M. Mannon of Company D on October 8, 1863, Private Henry Mason of Company G on January 17, 1864, Private Walker W. Brashears of Company L on May 10, 1864, and Quartermaster Sergeant J. B. Stark of Company F on February 9, 1865. Four months after the wedding, Mason was killed accidentally when a howitzer burst.

Many men died in the field from disease. The immunity of isolated farm boys to disease was poor, and conditions must have been difficult, indeed, for young men died in large numbers. The records show that the causes of death were such things as pneumonia, smallpox, "brain fever," dropsy, measles, pulmonary consumption, typhoid fever, or simply "disease."

Officers were permitted to resign, and enlisted men were discharged, due to various ailments which could not be cured. Major Ezra Fitch resigned due to disability on May 30, 1864. A year later, on May 1, 1865, Major Charles Galloway resigned because of rheumatism. In December 1863 Major James J. Johnson left the service due to poor health and the loss of hearing resulting from artillery fire.

One of the very first to go was First Lieutenant Frederick Ohlenmacker of Company F. He enlisted in Springfield, Missouri, in July 1862 at the age of only 21, but by October he wanted out. He submitted his resignation "for the reason that I am unhealthy and entirely unable to discharge the duties of my position." First Lieutenant George Reed, 32, felt in December 1862 that he could not go on "because of a disease in one leg which renders me unfit for duty."

First Lieutenant John W. Morris of Company H, though only in his late twenties, submitted his resignation on January 22, 1863. "My reasons for doing so," he explained in a letter to Colonel Harrison, "are want of physical capacity to undergo the exposure incident to a life in the field and a disease of the hip joint which at present entirely prevents me doing duty." Dr. Maynard examined him and endorsed his approval of the request, which was granted.

First Lieutenant Phillip Slaughter of Company E resigned due to "ill health, which unfits me for performing the duties of a soldier." Second Lieutenant John H. Thomas of Company G left due to pre-war injuries which "render me unable to endure camp life and the fatigues and exposures incident upon the profession of a soldier." Second Lieutenant Thomas J. Rice of Company M, age 23, resigned because "my health has failed and I have no hope of recovery."

Some men were discharged for non-medical reasons. Several young men were "claimed by their parents" because they were under age. An honorable discharge was given to First Lieutenant George S. Albright of Company L because one brother had been killed, another was in the service, and his aged parents were in need of care.

One member of the First Arkansas Cavalry was also a member of the Missouri Legislature. Private Patrick C. Berry enlisted in Stone County, Missouri, in July 1862, and was promoted to Quartermaster Sergeant of Company E. He was elected to the legislature, and Lt. Col. Bishop, in the absence of Colonel Harrison, gave him leave to attend the sessions. For reasons not recorded and now lost, Harrison very much disliked this soldier being in the legislature. He had earlier denied leave for attendance to legislative duties, and General Herron had agreed, but Bishop was more amenable. Upon his return from the leave granted by Bishop, Berry was "reduced to the ranks" by Harrison.

This dispute arose every time the Missouri legislature scheduled a session. Harrison simply would not budge. At one time Berry was called AWOL; another time a deserter. Finally, Secretary of War Edwin M. Stanton personally got involved. He sent a telegram specifically addressing Private Berry's situation on November 27, 1864. "Leave of absences always granted by the Department [of War]," he wired, "to members of the Legislature." Private Wilson Risley of Company B was simply discharged from the army when he was elected to the legislature.

A number of fatal accidents occurred in the regiment during the war. Jesse Long became perhaps the first fatality of the First Arkansas Cavalry when he was accidentally shot in the knee near Cassville, Missouri. He died on July 1, 1862.

Some sort of terrible accident occurred on May 17, 1863, at Crane Creek, Missouri. The records show only that Privates Joshua Steward and Marvin Steward of Company I, probably brothers, were both killed accidentally by Private James Adair.

While on prison guard duty Corporal Larkin Snow of Company K was killed by a gun discharging. On January 5, 1864, Private James Evans of Company L was killed by an accidental gunshot. Corporal Henry Mason of the howitzer section was killed in May 1864 when a gun burst. On May 29, 1864, while returning from a scout, Private James Stacey of Company L was fatally "shot from revolver in his own hands.... He was a good soldier, brave, obedient & respectful."

Lieutenant Joseph S. Robb was accidentally shot in the leg and bled to death. An obviously distressed Colonel Harrison wrote to the department adjutant on Christmas Eve 1863. "I would most respectfully report that Capt. Joseph S. Robb Co. L 1st Ark Cav died in hospital at this place at two o'clock a.m. 20th instant of accidental gunshot wound, ball passing between bones of the leg below the left knee cutting the main artery, causing excessive hemorrhage," Harrison wrote. "Very little is known of his relations except that his mother is a Mrs. Cynthia House of Palmyra, Iowa. Capt. Robb was a very gallant officer and tendered much important service to this government.... I believe him to have been thoroughly honest." Captain Robb's grave is in the Fayetteville National Cemetery.

Private Francis Brown of Company D died at his home near Fayetteville on February 15, 1863. Without disclosing the cause, the record-keeper noted with a cruel sense of humor that Brown "died of want of brains."

A number of men came back after the war in an effort to straighten out their service records. During the years from 1870 to 1910 men would apply to have their records changed to reflect that they had not in fact been AWOL or deserted. Regimental surgeon Henry J. Maynard, as an officer, was discharged for being AWOL. Years later he discovered this and applied to have the records changed on the grounds that he was not AWOL, but permitted to go home for being sick. While he was gone, the war ended and the regiment was mustered out. The petition was granted.

Private Hiram Teague was listed as having deserted at Flat Creek, Missouri, on March 1, 1863. Someone learned of this afterward and, knowing Teague's true fate, had the record changed in 1877. There was inserted into the file a statement that Teague was "sent upon secret service by Genl. Schofield was captured by the enemy on or about Mar 1, 1863, and hung by the Rebels the same night."

The First Arkansas Union Cavalry

Chapter 7
The Post Colony System

Despite the fact that the Federals were firmly in control of Fayetteville and could not be driven out, and that they could usually drive their enemies before them in any military confrontation, the fact of the matter was that northwest Arkansas was increasingly under Confederate dominion. Fayetteville was a Federal outpost surrounded by a Rebel-controlled countryside.

In late April of 1864 Major John Cosgrove of the Missouri State Militia 8th Union Cavalry noticed the Rebel upsurge. "I found the inhabitants of Benton County, Ark., and McDonald County, Mo.," he wrote, "to be the most disloyal I have seen since 1861, disposed to give all the aid and comfort in their power to the rebellion."

Lt. Col. Bishop of the First Arkansas Union Cavalry could clearly see the same phenomenon. On August 24th, Bishop reported after a scout, "Though we discovered no force of any consequence, there is still much work to be done in Northwestern Arkansas." A week later he wrote, stating, "There is scarcely a Union man to be found farming in the western portions of Benton and Washington Counties."

Colonel Harrison felt the same way. "On my return to this place," he wrote from Fayetteville on August 24th, "I found the rebels in such numbers and so insolent in this vicinity that I did not deem it prudent to send Captain H[ughes] forward at once." Then on November 10th, he reported, "These [guerilla] bands during the summer have given Union citizens great annoyance, constantly plundering and driving them from their homes, until the rebel rule in the surrounding country has been for a time almost complete." This situation caused a number of Union people to flee to Fayetteville for safety, where they became a burden to Harrison to shelter, feed and clothe.

Colonel Harrison approached this problem with three solutions. First, he waged a relentless war on Rebel-held areas by driving out anyone who would not take an oath of allegiance to the United States. Second, he adopted a policy of destroying the economic structure of Rebel areas. Third, he devised a "post colony system" for re-settling Union families in the hostile countryside. The oath requirement and the campaign of destroying mills was nothing new and was common throughout the country. The post colony system, however, was an unprecedented effort by a single Union regiment to make political, economic and social policy.

In the summer of 1864 Harrison launched his effort to re-structure northwest Arkansas society with a new and more stringent loyalty oath requirement. The oath became a vehicle by which freedoms, supplies and other benefits would be dispensed or withheld. On June 16th, the following order was issued:

> *I. All citizens of this place, male or female, who have not yet taken the Oath of Allegiance to the United States Government, and who shall fail to do the same before the 23d inst., will be sent beyond the limits of this command, except the two following classes; viz:*
> *- Married women, whose husbands are known to be loyal.*
> *- Children, who are minors.*
> *II. Ladies, whose husbands, or natural protectors, are in the service of the enemy north of the Arkansas River, will be sent beyond the lines.*
> *III. Transportation to Fort Smith will be furnished to such as do not desire, or are not allowed by the terms of Paragraph II, to take the Oath of Allegiance.*

The problem in northwest Arkansas in late 1864 was that the foreseeable future consisted of an endless guerilla war. The First Arkansas Union Cavalry could occupy Fayetteville and any ground where they happened to physically be, but as soon as they were gone so was Union control. The mountains, valleys, forests and caves of the area afforded to bushwhackers an advantage which no army of any size could ever completely neutralize; certainly one regiment could not do it.

While pursuing the policy of exiling anyone who would not take the loyalty oath, Harrison instituted a program of destroying the economic structure of the Rebel-controlled areas. This largely involved the destruction of mills and tanneries. At this time in history, mills were of extreme importance to local economies. They ground crops to food, produced furniture, cloth and tools. The loss of a mill was a terrible blow to nearby people. "The disabling of mills," Harrison wrote, "causes more writhing among bushwhackers than any other mode of attack; but they threaten to stay and fight me on boiled acorns."

Colonel Harrison's Letter Explaining The Post Colonies

Fayetteville Ark., December 23rd 1864
Col. A. W. Bishop, Adj't Gen'l of Ark.

Col: Today I replied by telegram [to your's of] 21st inst from Cairo. I write this as a simple memorandum to guide you in your entreaties for the suffering women and children of N. W. Arkansas. There are thousands of old men, women and children left here yet. You know their condition. I have from time to time worked to assist and protect them. Since you left I have established at their request post colonies at Rhea's Mill, Engle's Mill, Bentonville, Pea Ridge, Elm Springs and Huntsville, and am about organizing others at Mudtown, Mt. Comfort, Oreford Bend, Richland, McGuire's, West Fork and Hog Eye.

The plan is -
I. Fifty men capable of bearing arms unite and ask to be organized into a home guard company, and permission to settle on a large tract of abandoned land, which is all in one body.
II. They are organized, armed and move their families to the place.
III. They build a block house or small fort on the best point on the land (selected by me).
IV. They sign articles, agreeing to be loyal to the U.S. authorities; to abide by the laws and orders from the nearest military post; the laws and present [i.e., Union] Constitution of Arkansas, the proclamation of the President, etc. and are mustered in as Home Guards. They also agree to parcel out the land by vote, giving to each one all he wants to cultivate, but to have nothing in common, except common defence and obedience to law. Then all persons within 10 miles of these settlements are expected to enroll their names and belong to them, and none but rebels have so far objected.

Six of these settlements have made such progress, that each one will raise 10,000 bu. corn next season, and the Union Valley Settlement has agreed to deliver 1,000 tons of hay next season, if needed.

Bentonville and Elm Springs are swarming with people, who have moved in. Winningham is going to settle Mudtown with 50 Ark families returned from Mo.

All this is no chimera; it is half accomplished now, and the other companies are forming and will be at work within 10 days. Some of the forts are nearly done. The refugees have nearly all left this place and gone to the colonies. Bushwhackers' families have, 2/3 of them, been scared to Texas, and the others are fast going. No bushwhackers can exist here next season, if Gov't gives the poor people a little protection. Our regiment here is enough....

All are well, Yours for Arkansas, M. La Rue Harrison, Col. 1st Ark. Cav.

These policies, plus the war-weariness of the people, led many citizens to give up and leave the area. Large tracts of land became vacant, and in this Colonel Harrison saw the opportunity to win the war by re-settling the countryside with loyal people. If the people could not be won over, then they would be driven away and the area would be re-populated with Unionists.

The post colonies established as best they could independent and self-sufficient communities. The fact that colonists were illegally upon the land was conveniently ignored. They would elect a captain, establish a system of self-defense, plant crops, operate a blacksmith shop, and start a school. Each one generally had a healthy mix of men, women and children.

The concept of a system of Unionist colonization in a seceded state, devised and implemented by a single regiment, was unique. In so doing, Colonel Harrison far over-stepped the bounds of simple military duty. He delved into the arena of social and economic re-structuring, and somehow got away with it. Had some Union regiment in Virginia attempted such a thing, it surely would have been squelched by government authorities. But in the largely ignored backwash of the Trans-Mississippi it was done.

On February 18, 1865, the Federal outpost at Fayetteville was placed within the command of Brigadier General Cyrus Bussey. He was immediately beseeched by people complaining about the First Arkansas Union Cavalry. Two years of military occupation in one place by the same regiment inevitably led to making many enemies, and they made their presence known to Bussey. "It has been represented to me," Bussey wrote angrily to Harrison on the 28th, "that portions of your command have been committing the most outrageous excesses, robbing and burning houses indiscriminately. This must cease at once, and property of the people must be respected.... Let war be made on guerrillas and not women and children." Harrison wrote back to Bussey explaining his colony system, which pacified him for a period of time.

During the spring, however, Bussey continued to get complaints about the post colony system from people he considered to be loyal and of good character. Harrison, he was told, was compelling every male over fourteen to "join a colony or be considered a bushwhacker and suffer accordingly." He was also told that the people "have been living under a reign of terror for some time and that the colony system under

Harrison's compulsory order is oppressive." Bussey ordered Harrison to rescind his compulsory order and directed the Colonel to report to Union Governor Isaac Murphy to explain himself.

When Harrison received approval of the colonies from Little Rock, Bussey was not deterred. He shot off a heated letter on May 9th:

> Permit me to state that these colonies are not formed by the people, but by Colonel Harrison, who has virtually driven the people from their homes to these colonies. The people are very much opposed to the manner in which these colonies are organized, and hundreds of them have appealed to me for relief, stating that they did not want to leave their homes....
> At a public meeting in Fayetteville, Major Worthington, now dead, declared in a speech that any man who did not go into these colonies would be shot and have his house burned, etc. Colonel Harrison was present at this meeting, and did not correct the impression that went out - that every man must go into the colonies or be considered a bushwhacker....
> I have carefully investigated the facts, and have the testimony of nearly all the officers of the First Arkansas Cavalry, and many citizens who are vouched as loyal men, and they all express the same view of the subject. I have every reason to believe that Colonel Harrison is organizing these colonies for the purpose of controlling the vote of seven counties to elect him to Congress next fall.... Colonel Harrison has been over two years at Fayetteville, and I am convinced he has been there too long.... I have directed Colonel Harrison to permit the people to organize colonies, but not to interfere or compel them....

It is hard to imagine that the system instituted by M. La Rue Harrison was as harsh as Bussey maintains. His generally mild and forgiving ways with his own regiment portray a different personality. His professions of fondness for the Union people of Arkansas, and his devotion of three years of his life to their betterment, also indicate a genuine concern for them. Undoubtedly the rules were strict and probably harshly applied at times by subordinates, and there was a lack of appreciation for their enforcement by those who opposed the system, but it cannot be doubted that Colonel Harrison sought what he truly believed was best. Furthermore, Governor Isaac Murphy, another man of moderation, agreed. Murphy approved of the system and it was expanded to other areas of the state.

Fortunately for all, however, the end of the war was at hand, and all such issues were soon of no importance.

Chapter 8
Peace from the East

"During this year," Lt. Col. Bishop wrote of 1865, "a relentless warfare was carried on against the small band of guerillas who infested northwestern Arkansas, and many were killed." This was indeed so, for although the war was slowly moving to a foreseeable conclusion in the eastern part of the Confederacy, it continued unabated in the West without prospect for an end. The truth of it was the war could not be won by the Union Army in northwest Arkansas. The vastness of wild hills and mountains, the innumerable valleys and caves, and the unwillingness of so many of the people to submit to Federal authority rendered the task of military conquest virtually impossible without an overwhelming commitment of troops. If peace was to come, it would have to come from the East.

The first day of the year was a dark one for the First Arkansas Union Cavalry. Corporal John Rame and Private Isaac Bodine of Company E and Private John Fraley of Company B had been captured in the last few days of 1864 by guerillas. On New Year's Day they were executed.

The rest of January was costly as well. On the 6th Private William Littrell of Company A died of wounds from a skirmish at Hunstville. Ten days later Private William Ragsdale of Company C was killed by bushwhackers in Madison County.

On January 24th there was a sharp fight near Fayetteville. Francis McDonough and Calvin Smith, both corporals of Company I, and Private Richard Smith of the same company, were killed by guerillas. A similar incident occurred a month later. On the 17th of February a patrol got into a skirmish in Carroll County. Privates William Simpson and James R. Taylor of Company K were killed. Private James Carrigan of Company B "surrendered to the rebels Feb 19/65 & immediately shot."

The end of the war was nowhere in sight in northwest Arkansas.

It was a blow to the regiment when, less than a month before General Lee would surrender at Appomattox, Major Worthington was killed "at the head of his men while leading a charge against a column of bushwhackers." "It becomes my unpleasant duty to inform you that Major John I. Worthington of my Regiment was killed by a rifle ball," Colonel Harrison wrote to headquarters on March 15, 1865, "in an engagement with guerillas near Kingston Ark. on Sunday, March 12th 1865. His remains were interred with military honors at this place yesterday. He had lately been promoted from Captain of Co 'H.'"

Just two days later, the Confederates lost one of their most prominent leaders. Major William "Buck" Brown who had so persistently dogged the First Arkansas Union Cavalry, had his last skirmish. In Benton County there was another of those innumerable fire fights between the two old foes. Of the four bushwhackers killed, one of them was Brown. There was no sorrow about this in the camp of the First Arkansas Union Cavalry.

Another two days passed and Corporal Samuel Hood of Company A was on his last patrol. He was killed on March 17th by guerillas in Washington County.

Within the regiment a controversial situation arose in March when Sergeant John P. Todd of Company M was ordered arrested to be executed for murder. This was a continuation of the story of the murder of Lieutenant James Roseman in June 1864. Private Joseph Wisdom had murdered Roseman, then escaped and joined the enemy. He was active as a guerila until killed in the early spring of 1865.

The murderer had a brother, Creed Wisdom, who stayed on in the First Arkansas Union Cavalry. Sgt. Todd was ordered with several others to keep an eye on Creed, who was suspected of disloyalty. On July 7, 1864, Wisdom and Private Eli Bell were caught in the act of stealing ammunition. Todd and his men confronted them, Bell was killed and Creed escaped. He was thought to have joined his brother for a time, but was ultimately picked up as a deserter and brought back to the regiment.

Creed frequently said that he would kill Todd, and on September 10, 1864, there was a confrontation between the two. To save his own life, Todd killed Creed Wisdom. A confident defense was astonished when Todd was convicted of murder and ordered to be executed by a firing squad. The order was ignored by Colonel Harrison and Todd was returned to his normal duties for several months.

When the First Arkansas came under the jurisdiction of General Cyrus Bussey, he learned of the situation and ordered Harrison to arrest Todd and have him executed. A flurry of activity was generated at the regimental level to stop it. Sergeant Doctor B. Norris of Company M wrote an affidavit that he heard Creed Wisdom threaten Todd's life.

Nearly every officer in the regiment wrote a letter to Governor Murphy on March 7, 1865, pleading for Todd. "We know well the reputation of Sergeant Todd as a citizen & a soldier," they wrote. "We know him to be honest, honorable and patriotic. We believe that Wisdom, who was killed by him, was a Rebel bushwhacker and a spy." They concluded by saying that they "most earnestly petition & request that you use your influence with the Major Genl Commanding the Dept for a revocation of the order for his execution."

On March 17th Colonel Harrison wrote his own letter. "Todd is one of those few southern men who has been faithful to the national cause ever since the commencement of the rebellion...," he wrote. He recited that he was honest and moral, a good family man and a gentleman, and that he never turned his back on the enemy and had personally killed Buck Brown and other Rebels. "I must earnestly and respectfully entreat and petition in the name of all the officers and enlisted men of my regiment that you will revoke the sentence of death," he concluded.

Based on these entreaties, General Bussey reversed himself. He joined in the petitioning by asking not for a commutation of the sentence but for an outright pardon. Not only was Todd then pardoned, he was promoted. On June 4th he was made a 2nd Lieutenant in Company F.

While the status quo of endless war persisted in northwest Arkansas, matters were evolving quickly in the East. With the war winding down, Colonel Marcus LaRue Harrison was breveted to the temporary rank of Brigadier General. It was an honor he had earned, and he was always proud of it. On April 2nd, Richmond fell and a week later Robert E. Lee surrendered the Army of Northern Virginia at Appomattox. The news was soon relayed to the Trans-Mississippi. The Federal troops and post colonies fired guns in celebration.

For the Confederate bands in the hills, though, it took time to decide what to do. Should they surrender or keep on fighting? While they thought his over, the pointless killings continued. On April 12th Corporal Ira Williams of Company F died of wounds received in a skirmish near

Bentonville. On May 1st, Private John Millsap, who had twice deserted, was wounded in action at Huntsville. First Lieutenant Warren Munday was wounded near Bentonville.

Guerillas kept operating not only through April but into May. Twenty-year-old Isaac Watkins of Company E had enlisted in Fayetteville in October of 1863. He was wounded there on December 27, 1864, but by May 16, 1865, he was back on duty. That was the day he was killed by Rebel bushwhackers. He was the last sacrifice required of the First Arkansas Union Cavalry upon the altar of civil war.

Finally, on June 2, 1865, Confederate General E. Kirby Smith surrendered the Department of the Trans-Mississippi. The Civil War in Arkansas was over. On August 23, 1865, the First Arkansas Union Cavalry was mustered out of the service of the United States. Of the 1,765 men who had served in the regiment, about 110 (6.2%) had been killed in action or died of battle wounds, and approximately 235 (13.8%) died of diseases and accidents. No systematic records now exist of how many were wounded, although about fifty can be documented by name. Statistically, however, it is likely that about three times as many were wounded as killed. Over the course of the war, this would have been an estimated 300 to 350 men wounded. By the end of the war, then, one in five men who had served in the regiment was dead. Another one in five was wounded. At forty per cent dead or wounded, no man could feel that he had an easy service in the First Arkansas Union Cavalry.

The regiment had been asked to perform a very difficult task - to suppress a rebellion in its home district. Probably no other Southern regiment in the Union Army had the assignment not only of fighting its own neighbors, but then occupying its home ground for nearly three years. The land area in which this single regiment had primary responsibility was as large as the entire northern Virginia battleground between Washington and Richmond.

There are two memorials to the First Arkansas Union Cavalry today. One is Headquarters House in Fayetteville, which serves as a continuing reminder that these men were once here. The most important memorial, however, the one that its soldiers most wanted, is in Little Rock at the State Capitol. There, now unchallenged, the flag of the United States flutters in the wind over Arkansas.

THE END.

Endnotes

1. Woods, James M., *Rebellion and Realignment: Arkansas's Road to Secession* (Fayetteville, University of Arkansas Press, 1987), p. 160.

2. Goodspeed, ed. *History of Benton, Washington, Carroll, Madison, Crawford, Franklin and Sebastian Counties, Arkansas,* (Chicago: 1889), p. 227-232.

3. Bishop, Albert W., *Loyalty on the Frontier, or Sketches of Union Men of the South-West* (St. Louis, E. P. Studley and Co., 1863)., pp. 88-104.

4. Bishop, pp. 11-12.

5. Bishop, Albert W., *An Oration Delivered at Fayetteville, Arkansas,* July 4, 1865.

6. *Official Records of the War of the Rebellion*, Series 3, Volume 2, p. 958.

7. Bishop, pp. 62-63.

8. *Official Records*, Series 1, Volume 22, Part 1, p. 72

9. Regimental Records, Union Regiments from Arkansas, First Regiment of Cavalry, under the name of Marcus LaRue Harrison, National Archives Microfilm.

10. *Official Records*, Series I, Volume 22, Part 1, p. 137.

11. Tillie, Nannie M., ed, Federal on the Frontier: The Diary of Benajmin F. McIntyre (Austin, 1963), p. 58.

12. *Official Records*, Series 1, Volume 22, Part 1, pp. 102-103.

13. Regimental Records, Union Regiments from Arkansas, First Regiment of Cavalry, under the Regimental Returns, the

Company M Returns, and the Company F Returns, National Archives Microfilm.

14. *Official Records*, Series 1, Volume 22, Part 2, p. 801.

15. *Official Records*, Series 1, Volume 22, Part 1, p. 312.

16. *Official Records*, Series 1, Volume 22, Part 2, p. 848.

17. *Official Records*, Series 1, Volume 22, Part 2, p. 192.

18. *Official Records*, Series 1, Volume 22, Part 2, p. 191.

19. Returns of U.S. Military Posts, 1800-1916, Fayetteville, Arkansas, April 1863.

20. Regimental Records, Union Regiments from Arkansas, First Regiment of Infantry, under the name of Mathew W. Sumner, National Archives Microfilm.

21. *Official Records*, Series 1, Volume 22, Part 1, p.306. Regimental Records, Union Regiments from Arkansas, First Regiment of Cavalry, under the name of Marcus LaRue Harrison, National Archives Microfilm.

22. Bishop, Albert W., Battle Report, April 22, 1863, Regimental Records for the First Arkansas Union Cavalry, National Archives in Washington, D.C.

23. *Official Records*, Series 1, Volume 22, Part 1, pp. 306, 310.

24. Regimental Records, Union Regiments from Arkansas, First Regiment of Infantry under the names of Francis W. Cannon and Gilbert C. Luper, and the First Regiment of Cavalry, under the names of Thomas Bingham,. Cyrus Barber and Marcus LaRue Harrison, National Archives Microfilm.

25. Regimental Records, Union Regiments from Arkansas, First Regiment of Cavalry, under the name of Marcus LaRue Harrison, National Archives Microfilm.

26. Regimental Records, Union Regiments from Arkansas, First Regiment of Cavalry, under the names of William Johnson, Davis Chyle and Doctor B. Norris, National Archives Microfilm.

27. *Official Records*, Series 1, Volume 22, Part 1, p. 312.

28. Baxter, William, *Pea Ridge and Prairie Grove* (Cincinnati, Hitchcock & Walden, 1869), pp. 225-226.

29. Regimental Records, Union Regiments from Arkansas, First Regiment of Cavalry, under the name of Marcus LaRue Harrison, National Archives Microfilm.

30. Walker, Edwin S., Editor, *Genealogical Notes of the Carpenter Family*, (Springfield: 1907), pp. 142-143.

31. Bishop, Albert W., battle report, April 22, 1863, Regimental Records for the First Arkansas Union Cavalry, National Archives in Washington, D.C.

32. *Official Records*, Series 1, Volume 22, Part 1, p. 311.

33. *Official Records*, Series 1, Volume 22, Part 1, p. 311.

34. *Official Records*, Series 1, Volume 22, Part 1, p. 305.

35. *Official Records*, Series 1, Volume 22, Part 1, pp. 306-308.

36. *Official Records*, Series 1, Volume 22, Part 1, pp. 306-308.

37. *Official Records*, Series 1, Volume 22, Part 1, p. 309.

38. *Official Records*, Series 1, Volume 22, Part 1, p. 310.

39. Bishop, p. 217.

40. Little Rock *True Democrat,* April 29, 1863.

41. *Official Records*, Series 1, Volume 22, Part 1, p. 311.

42. *Official Records*, Series 1, Volume 22, Part 2, p. 246.

Selected Bibliography

Allen, Desmond Walls, *First Arkansas Union Cavalry* (Conway, Arkansas, 1987).

Anderson, John Q., Editor, *Campaigning with Parson's Texas Cavalry Brigade, CSA: The War Journals and Letters of the Four Orr Brothers, 12th Texas Cavalry Regiment* (Waco, 1967).

Baxter, William, *Pea Ridge and Prairie Grove, or Scenes and Incidents of the War in Arkansas* (Cincinnati, 1864).

Bishop, Albert, *Loyalty on the Frontier, or Sketches of Union Men of the South-West* (St. Louis, 1863).

Bishop, Albert, *Report of the Adjutant General*, 1867.

Bishop, Albert, *An Oration Delivered at Fayetteville, Arkansas, July 4, 1865* (New York, 1865).

Campbell, William S., *One Hundred Years of Fayetteville, 1828-1928.* (Fayetteville, 1977).

Harrell, John M., *Confederate Military History, Volume 14, Arkansas* (Wilmington, 1988).

Harrison, Elizur B., "The Battle of Fayetteville," *Flashback*, Washington County Historical Society, Volume 18, No. 2, (May 1968).

Little Rock True Democrat, April 29, 1863

National Archives Records for (1) The First Arkansas Union Cavalry, (2) The First Arkansas Union Infantry. (3) Monroe's First Arkansas Confederate Cavalry, (4) Carroll's First Arkansas Confederate Cavalry (listed under Gordon's Cavalry, (5) Miscellaneous Arkansas Records, (6) Dorsey's Missouri Cavalry, (7) The 12th Texas Cavalry, (8) The 18th Texas Cavalry, and (9) Post reports. On microfilm and at the Washington D.C. office.

U.S. War Department, *The War of the Rebellion: A compilation of the*

Official Records of the Union and Confederate Armies, 128 volumes.

Walker, Edwin S., *Genealogical Notes of the Carpenter Family*, (Springfield, Illinois, 1907).

Yeater, Sarah J., "My Experiences During the War Between the States," *The Arkansas Historical Quarterly*, Vol. IV, No. 1, (Spring 1945).

Washington Telegraph, April 29, 1863.

Index of Members of the First Arkansas Union Cavalry

Adair, James - 86
Albright, George S. - 85
Allison, James F. - 80
Amos, Lucien - 33, 51
Asbill, John - 51
Bage, Abednego - 27
Bage, John A. - 27
Bar, Hill B. - 57
Barber, Cyrus - 37, 49, 81
Bays, Jerome - 49
Bell, Eli - 96
Bell, James D. - 37, 51
Bell, Joseph - 49
Berry, Patrick C. - 85-86
Bishop, Albert W. - 4-6, 8, 9, 11, 12, 15, 21, 25, 36, 38, 40, 43, 45-47, 49, 63, 67, 76, 77, 79, 84, 85, 89, 91, 95
Black, John G. - 32
Blevins, Elias - 49, 51
Bloyed, William - 23
Bodine, Isaac - 95
Boman, Wiliam - 64
Bonine, John - 7
Botefuhr, Hugo C. - 36, 68, 72
Brashears, Walker W. - 84
Bridges, John S. - 54
Brooks, Thomas - 56
Brown, Francis - 86
Burrow, William M. - 36, 51
Burrows, Reuben - 51
Byrd, George - 67
Caffee, Amos - 38, 40
Carrigan, James R. - 95
Carrigan, John - 26
Caughman, Nelson - 64
Cavin, James - 71-72
Center, Guilford - 19
Chyle, Davis - 23, 39, 51
Connor, Richard - 56
Cook, Hugh - 45
Cooper, Benjamin - 67
Crawford, John B. - 61
Crawford, Thomas - 64
Daniels, James - 9
Davis, John - 83
Davis, William F. - 51
Day, Joshua W. - 20
DePriest, John - 77-78
Dienst, John - 32
Dillingham, George W. - 61
Dotson, Bloomington - 67
Downey, John R. - 67
Dunfield, Ahaz - 9
Dunnell, Cahrles - 65
Echols, James - 20
Evans, James M. - 65, 86
Fears, Josiah - 51
Fitch, Ezra - 36, 38, 40-44, 47, 77, 84
Forehand, John - 61
Fraley, John - 95
Freeman, Jesse - 63
Galloway, Charles - 6, 9, 25, 84
Gardiner, Jacob - 65
Gilbert, Isaac T. - 65
Gildemeister, Henry W. - 7
Gilstrap, Jesse - 10, 56, 79, 80
Gilstrap, Thomas - 5
Grady, Jasper - 64
Graham, Benjamin K. - 45, 51
Gregg, Allen C. - 51

Grubb, John A. - 51
Hale, Elisha - 27
Haley, S. D. - 44-45
Harp, John A. - 51
Harper, John - 27
Harrison, Edward M. - 6
Harrison, Elizur B. - 6, 34, 35, 38, 68
Harrison, Marcus LaRue - 4, 5, 11-17, 25, 27, 30-38, 40-49, 53-55, 57-61, 65-69, 71-78, 82, 85-86, 89-97
Hayes, John - 51
Hendricks, Larkin - 61
Herridan, George - 64
Hilbert, Mathew W. - 76-77
Hixon, George - 64
Hixon, William - 67
Hobbs, James - 27
Hodges, Anson - 20
Holmesly, Randolph - 61
Hood, Samuel - 96
Hopkins, DeWitt C. - 37, 47, 68
Hottanhour, Gustavus - 51
Howry, George W. - 55
Hughes, George - 26
Hughes, James - 26
Hunt, Thomas J. - 5, 36, 47, 59-61, 65, 76, 80, 82
Hunter, John - 64
Hutchison, Alred - 77-78
Jack, James - 51
James, John - 64
Jernigan, Lawson D. - 80-81
Johnson, Herod - 67
Johnson, James J. - 6, 27, 84
Johnson, John K. - 62
Johnson, Micajah - 27
Johnson, William O. - 26
Johnson, William S. - 38, 39
Jones, Charles Madison - 74
Jones, J. M. - 27
Jones, M. V. - 61
Keiser, Jacob H. - 78
Kelly, Robert B. - 26
Kimes, Francis M. - 67
Kincaid, Elam O. - 78
King, George R. - 65
Kise, Frederick - 51
Lewis, Bracken - 75
Lewis, Henry C. - 51
Littrell, Joseph G. - 63
Littrell, Joseph N. - 63
Littrell, Samuel - 11, 63
Littrell, William - 63, 95
Long, Jesse - 86
Lowe, William - 72-73
Luna, John T. - 64
Mack, Robert - 27
Mack, Rowen E. - 11, 36, 64
Manes, Jesse - 51
Mannon, Francis M. - 85
Mannon, William - 65
Marshall, William - 67
Mason, Henry - 84, 86
Maynard, Henry - 77, 78, 85, 87
Maynard, Willis - 81
McDonough, Francis - 95
McGlothlin, Burley - 64
McGuire, Moses - 64
McLaughlin, William - 64, 83
Means, William - 64
Merrill, Joseph - 27
Messenger, Daniel - 27
Messenger, William L. - 37
Miles, Frederick - 20
Miles, John - 27
Miller, John - 83
Miller, William - 51
Mills, Enos - 58
Millsap, John - 83, 98
Moore, Curry - 13
Morris, George A. - 51

Morris, John W. - 85
Morrison, Hamilton - 65
Moton, Theodore - 65
Munday, Warren W. - 81, 98
Nail, Jesse - 51
Nelson, George - 13
Norris, Alvin - 81
Norris, Doctor B. - 39, 51, 97
Norris, Jesse - 56
North, Reuben - 84
Ohlenmacher, Frederick - 84
O'Neal, Charles B. - 20
Owens, Willis - 65
Oxford, Jacob - 51
Packer, Jeremiah P. - 66
Parker, Elijah - 67
Patrick, William - 26
Pearson, Albert - 6, 73
Peerson, W. C. - 7, 69-70
Pennington, Wesley - 67
Peters, Michael - 67
Phillips, Luther - 11
Pitts, Alfred - 27
Poor, Henry - 64
Quinton, William J. - 51
Ragsdale, William - 95
Rainey, John - 67
Rainey, Joseph - 81
Rame, John - 95
Reed, George - 85
Reed, John - 49
Reed, Joseph - 67
Reed, Robert - 61
Reel, Jacob J. - 7
Rice, Thomas J. - 85
Rickets, Joel - 65
Riddle, Jonas - 33, 51
Risley, Wilson - 86
Robb, Joseph S. - 7, 33, 40, 43, 45, 56, 57, 64, 86
Robinson, Ezekial - 20
Robinson, James G. - 60

Robinson, William F. - 20
Rose, Jesse - 61
Roseman, James - 6, 47, 62, 65, 66, 67, 83, 96
Ross, Hiram - 10
Russell, George W. - 33, 51
Rutherford, John - 51
Rutherford, Noel - 26
Scaggs, Hannibal - 51
Scott, Robert J. - 64
Shahan, Hiram - 14
Shibley, James - 64
Simpson, William - 95
Sisemore, George W. - 51
Slaughter, Phillip - 85
Smith, Calvin - 95
Smith, Richard - 95
Snow, Larkin - 86
Stacey, James - 86
Standlee, James - 11
Stanley, Joel - 64
Stark, Denton D. - 27, 64
Stark, J. B. - 84
Steward, Joshua - 51, 86
Steward, Marvin - 86
Stockton, David - 67
Stockton, George - 27
Stockton, John - 27
Stockton, Samuel - 27
Strickland, Levi - 51
Strohan, John - 13
Strong, Frank - 47
Taylor, James R. - 95
Taylor, Jordan - 51
Teague, Hiram - 87
Tefft, Jonathan E. - 40
Temple, Francis M. - 51, 84
Thomas, John H. - 85
Thomson, Robert - 25, 55
Todd, Elijah - 51
Todd, John P. - 96, 98
Travis, Robert E. - 25, 38

Turman, John - 32, 64, 65, 81
Vaughan, John - 25, 60
Vice, George - 61
Ward, Francis M. - 81
Watkins, Daniel - 23
Watkins, Isaac - 98
Weldon, James - 13
Whitcomb, Orin - 81
Wikle, Jared - 58
Wilburn, Francis D. - 69
Wilhite, Thomas - 2, 3, 5, 7, 25, 79-80
Williams, Ira - 97
Willis, Jacob - 64
Wills, Cyrus - 82
Wilson, - 65
Wilson, James H. - 7
Wilson, T. M. - 65
Wimpy, Richard H. - 32
Wisdom, Creed - 96, 97
Wisdom, Joseph - 67, 97
Wood, George - 65
Wooten, William - 51
Worthington, John I. - 6, 9, 10, 56, 61, 95, 96
Yoes, Jacob - 24
York, William J. - 51
Youngblood, Theodore - 11, 76

www.ingramcontent.com/pod-product-compliance
Lightning Source LLC
Chambersburg PA
CBHW060844050426
42453CB00008B/816